ASPECTS OF HOMOGENEOUS CATALYSIS

Volume 2

ASPECTS
OF HOMOGENEOUS
CATALYSIS

A Series of Advances

EDITED BY

RENATO UGO

ISTITUTO DI CHIMICA GENERALE ED INORGANICA
MILAN UNIVERSITY

VOLUME 2

D. REIDEL PUBLISHING COMPANY

DORDRECHT-HOLLAND / BOSTON-U.S.A.

CHEMISTRY

ISBN 90 277 0522 4

Published by D. Reidel Publishing Company
P.O. Box 17, Dordrecht, Holland

Sold and distributed in the U.S.A., Canada and Mexico
by D. Reidel Publishing Company, Inc.
306 Dartmouth Street, Boston, Mass. 02116, U.S.A.

Printed in Italy
Centro Grafico Linate di Rossi-Santi - San Donato Milanese

Editorial Board

Contents of Volume 2

Hydroformylation of Olefins with Carbonyl Derivates of the Noble Metals as Catalysts
László Markó

Addition of hydrogen cyanide to mono-olefins catalyzed by transition metal complexes
E. S. Brown

Nickel Catalyzed Syntheses of Methyl-Substituted Cyclic Olefins, an Example of Stepwise Carbon-Carbon Bond Formation Promoted by a Transition Metal Complex

P. Heimbach

Dimerization of Acrylic Compounds

Masanobu Hidai and *Akira Misono*

ASPECTS OF HOMOGENEOUS CATALYSIS

Volume 2

Chapter 1

Hydroformylation of Olefins with Carbonyl Derivates of the Nobles Metals as Catalysts

LÁSZLÓ MARKÓ

Institute of Organic Chemistry, University of Chemical Industries, Veszprém, Hungary

4

1. INTRODUCTION

Cobalt carbonyl was regarded for a long time as the only catalyst suitable for hydroformylation. This was ever the situation nearly a quarter of a century after the discovery of the reaction by Roelen in 1938, although industry has almost certainly devoted much effort to the search for other, more active or more selective catalysts. Work was mainly directed towards iron and nickel, the two neighbouring elements of cobalt in the periodic table, but in spite of numerous efforts, no practical results could be achieved.

The development of new catalysts was hampered mainly by the prejudice that the third neighbour of cobalt, rhodium is a "noble" metal. Noble metals were regarded at that time by those working in this field of chemistry as chemically unreactive and therefore not very interesting elements; the practical use of which would be prohibited anyway by economic considerations. As so often happens in science, it took considerable time to overcome this prejudice. However, soon after it was recognized that the complexes of noble metals have intriguing and useful catalytic properties, a breakthrough was achieved. This occurred around 1965 and the past five years have brought a rapid advance in the number of organic reactions catalyzed by noble metal complexes. One of the most spectacular achievements in this respect was the development of hydroformylation catalysts based on noble metals, especially the carbonyl complexes of rhodium.

Rhodium and iridium carbonyls as catalysts for hydroformylation were first mentioned in a patent of the Chemische Verwertungsgesellschaft Oberhausen m.b.H. [88] (inventor G. Schiller). This patent, which was disclosed in 1956 (and applied as early as 1952), correctly described the advantages of rhodium (the use of lower temperatures and pressures and its higher activity relative to cobalt), nevertheless it apparently remained unnoticed. It was evidently the patent of the Esso Research and Engineering [52] (inventor V.L. Hughes), disclosed in 1959, which received the necessary publicity and stimulated scientific research. The first publication in this field was the work of N. S. Imyanitov and D. M. Rudkovskii [54] in 1963, which was then followed by a large number of scientific contributions. The modification of the rhodium catalyst by using phosphine complexes of rhodium or by adding phospines to the reaction mixture was first achieved by G. Wilkinson's school [75] and the researchers of the Shell Oil [91] (L. H. Slaugh and R. D. Mullineaux) in 1965 and 1966, respectively. This area has received the most attention in recent years, since the great number of possible combinations offers a promising opportunity to find more active and more selective catalysts.

Earlier reviews appeared in 1968 [47, 108] only partly covering the subject. The present article covers the literature up to the end of 1971. An effort has been made to cover patents as far as possible. This is seldom done in review articles, partly because of the voluminous and at the same time inexact description of experimental material and partly because of the lack of availability of the patent data. However, experience often shows that patents may be good starting points for basic research. To reduce the amount of data, only the examples and not the claims were considered.

2. COMPARISON OF THE ACTIVITY OF GROUP VIII METALS

Before dealing separately with the noble metals as hydroformylation catalysts, their relative activities are compared in Tables 1-4. The data do not allow any quantitative evaluation, but some qualitative comparisons can be made as follows:

a) Rhodium is by far the most active metal for hydroformylation of monoolefins, surpassing in this respect even the "classical" metal, cobalt.

b) Iridium and ruthenium carbonyls are already significantly less active than cobalt, and $Os_3(CO)_{12}$ shows an even smaller activity.

c) The catalytic activity of platinum and palladium compounds in hydroformylation is even less, although it must be mentioned that in the case of these metals the experimental material is rather scanty (See Section 6.).

Table 1

HYDROFORMYLATION OF α-METHYLSTYRENE WITH COBALT-, RHODIUM- AND IRIDIUM-CARBONYLS AS CATALYSTS [54, 55]

Metals were used in the form of carbonyls; Catalyst concentration: 10 mg atom metal $\times 1^{-1}$; Pressure: 300 atm (1.2 H_2 + 1 CO)

Metal	Temperature, °C	Initial rate as pressure drop, atm. min^{-1}	Yield, %	
			aldehyde	isopropyl-benzene
Co	130	0.6	45	54
Rh	130	⩾ 450	99	3
Co	170	4.0	—	55
Ir	170	0.3	4	90

Table 2

HYDROFORMYLATION OF CYCLOHEXENE WITH COBALT-, RHODIUM- AND IRIDIUM-CARBONYLS AS CATALYSTS [56, 57]

Metals were used in the form of carbonyls; p_{H_2} = 250 atm; p_{CO} = 150 atm

Metal	Catalyst concentration, mg atom metal $\times 1^{-1}$	Temperature, °C	Conversion of cyclohexene, %	
			to aldehyde	total
Co	3	150	80	82
Rh	trace	150	97	95
Ir	3	150	2	10
Ir	3	200	15	98

Table 3

HYDROFORMYLATION OF PROPYLENE WITH COBALT-, RHODIUM-, RUTHENIUM- AND OSMIUM-CARBONYLS AS CATALYSTS [81]

Catalyst	Catalyst concentration, mg atom metal $\times 1^{-1}$	Temperature, $^{\circ}C$	p_{H_2} atm	p_{CO} atm	Reaction time, h	Conversion of olefin to carbonyl compounds, %	$n : i$ aldehyde ratio
$Co_2(CO)_8$	5.8	150	75	75	0.16	81.0	2.4
$Rh_4(CO)_{12}$ (*)	5.3	110	110	110	1	80.0	1.1
$Ru_3(CO)_{12}$	5.8	150	75	75	1	23.0	2.6
$Os_3(CO)_{12}$	18.5	180	80	80	2	74.0	1.0

(*) From Ref. [105].

Table 4

HYDROFORMYLATION OF PROPYLENE WITH COBALT-, RHODIUM-, IRIDIUM-, RUTHENIUM- AND OSMIUM-CARBONYLS AS CATALYSTS [82]

Catalyst	Catalyst concentration, mg atom metal $\times 1^{-1}$	Temperature, $^{\circ}C$	p_{H_2} atm	p_{CO} atm	Reaction time, hours	Aldehyde yield, %	Aldehyde $n : i$ ratio
$Co_2(CO)_8$	23.4	110	75	75	1	93.7	4.0
$Co_2(CO)_6(PBu_3)_2$	23.4	140	30	15	3	80.0	6.1
$Rh_4(CO)_{12}$ (*)	5.3	110	110	110	1	80.0	1.1
$Ir(CO)(PPh_3)_2Cl$ (**)	9.3	130	15	15	6	67.6	1.8
$Ru_3(CO)_{12}$	23.4	110	75	75	7	40.2	2.8
$Os_3(CO)_{12}$	18.5	180	80	80	2	74.0	1.0

(*) From Ref. [105].
(**) From Ref. [5].

An effort was made to compare the catalytic activity of simple rhodium- and cobalt-carbonyls in hydroformylation on a quantitative basis. Based on kinetic measurements an activity ratio of Rh : Co = 12.000 : 1 was obtained [46].

As can be seen from Tables 3 and 4, the normal : iso aldehyde ratio is usually smaller, in the case of the simple carbonyls of noble metals, than that which can be obtained with catalysts based on cobalt. Since this ratio is strongly influenced by changing the reaction conditions and by using phospines, etc., this aspect will be discussed in more detail later.

3. HYDROFORMYLATION WITH RHODIUM COMPLEXES AS CATALYSTS

By far the most work on hydroformylation with noble metals as catalysts has been done with rhodium, this metal showing the greatest activity. A charac- teristic of this work is the great variety of rhodium compounds and ligands (such as phosphines, phosphites, etc.) used as catalysts. Therefore a survey of the applied catalysts will be given first, this will then be followed by sections on the hydroformylation of different substrates and lastly by a discussion of the reaction mechanism.

3.1. Survey of Applied Catalysts and Catalyst Combinations

3.1.1. METALLIC RHODIUM, RHODIUM CARBONYLS AND SIMPLE RHODIUM COMPOUNDS

If rhodium alone is used as a catalyst, it is applied in the form of rhodium on a support [20, 21, 22, 40, 41, 42, 52, 70, 88]. The use of $Rh_4(CO)_{12}$ [1, 2, 45, 46, 48, 71, 81, 82, 87, 98, 99, 100, 105], $[Rh(CO)_2Cl]_2$ [25, 71], rhodium salts of carboxylic acids [34, 116], Rh_2O_3 [4, 18, 28, 29, 30, 31, 32, 52, 66, 72], $RhCl_3$ [71] or $RhCl_3 \cdot 3H_2O$ [44, 45, 77, 116]. has been described. Sometimes only terms such as "rhodium carbonyl" [40, 41, 42, 54, 55, 56, 57] "rhodium nitrate" [71, 115] and "rhodium chloride" [72] are used. "Rhodium carbonyl" is as- sumed to mean usually $Rh_6(CO)_{16}$, since this is the rhodium carbonyl which can be prepared and handled most easily. The term "rhodium trichloride" may mean either the anhydrous $RhCl_3$ or the "trihydrate" $RhCl_3 \cdot 3H_2O$ (the latter actually being a mixture of Rh (III) chloro and aquo complexes [58]), since both are transformed by carbon monoxide to rhodium carbonyls [12, 51] under the conditions generally used for hydroformylation. There is no indication that the use of any of the above mentioned rhodium sources would be more advantageous than the other. However, since infrared spectroscopic evidence suggests $Rh_4(CO)_{12}$ and $Rh_6(CO)_{16}$ to be the complexes principally present under conditions of hydroformylation [45, 48], these two forms of rhodium seem to be the most suitable to study as catalysts if no other ligands are to be used.

3.1.2. RHODIUM CATALYSTS MODIFIED WITH PHOSPHINES AND RELATED COMPOUNDS

The rôle of phosphines and other ligands in the formation of the catalytically active complexes and their effect on the rate, selectivity, etc. of the reaction will be discussed later in some detail. Here it should only be mentioned that these ligands can be applied either by using rhodium complexes already containing them (e.g. $RhH(CO) (PPh_3)_3$, $Rh(CO) (PPh_3)_2Cl$) or by adding them (usually in great excess with respect to rhodium) to the reaction mixture. Often both methods are used simultaneously.

Table 5 lists the Rh + P (As) catalyst combinations reported to date and gives the more important conditions of their use. Obviously, rhodium catalysts modified with phosphines and related ligands have found a wide range of applications and are at present the hydroformylation catalysts being most intensively investigated. A detailed discussion of these catalysts will follow in Sections 3.3-3.7. The great interest taken in phosphine-modified rhodium catalysts seems to be well justified if one considers some outstanding results achieved in this field. For example in catalytic hydroformylation at atmospheric pressure using $RhH(CO) (PPh_3)_3$ [11, 25] or $Rh(CO) (PR_3)_2(OOCR')$ [8, 62], $n : i$ aldehyde ratios as high as 25 : 1 have been obtained [11, 25] and the continous hydroformylation of propylene, without the necessity of catalyst recovery in a gas sparged reactor, has been achieved using $Rh(CO) (PPh_3)_2Cl$ [50].

3.2. Effect of Halogen Acceptors and Nitrogen Compounds

Although the methods used for rhodium carbonyl preparation from chlorine containing rhodium compounds [12, 13, 14, 51], usually also use some halogen acceptor, only in one case [91, 93] has the application of such an additive been reported when rhodium chloride was used as catalyst. Since rhodium chlorides, when used as catalysts, are always used under high pressure conditions in steel autoclaves, it is most probably the metal of the autoclave wall which serves as the halogen acceptor. The rôle of solvents (e.g. alcohols) cannot, however, be excluded.

The complex 1, 2, 6- $RhPy_3Cl_3$, which has also been used as a hydroformylation catalyst [75], already contains the halogen acceptor as a ligand.

With $Rh(CO) (PPh_3)_2Cl$ as catalyst it was shown [25], that the addition of triethylamine eliminates the induction period, which is normally always observed in the hydroformylation reaction, by neutralizing the HCl formed (Section 3.7.3.) Perhaps the same rôle can be ascribed to the N-containing ligands in $Rh(CO)_2LCl$ type complexes [37].

Table 5

RHODIUM CATALYSTS MODIFIED WITH PHOSPHINES, PHOSPHITES OR ARSINES

Rhodium compound	Added ligand	Ligand: Rh ratio	Temperature, °C	Pressure, atm	Reference
Rhodium on carbon	PBu_3 PPh_3 $P(OAr)_3$	0-27	90-110	5-170	84, 85
$Rh_6(CO)_{16}$	PPh_3	1.5	82	15-42	38
"$Rh_2(CO)_8$"	$P(OPh)_3$	2.5	140	200	23
Rh-ethylhexanoate	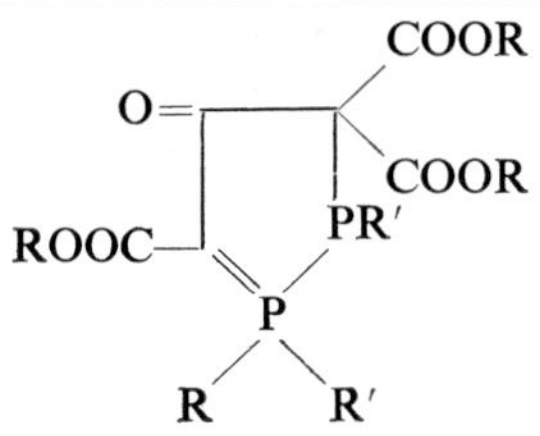 1-phenyl phospholine-2	—	—	—	34
Rh_2O_3	PR_3 PPh_3 $P(OEt)_3$	30-120	80-150	200	29, 35, 36
$[Rh(CO)_2Cl]_2$	PPh_3	0-50	100-150	35-70	17
$Rh(C_5H_5)_2Cl$	PPh_3	0-50	100-150	35-70	17
$RhCl_3$	PBu_3 (+ CH_3COONa)	2-8 (+6-8)	195	35	91, 93
	OPR_3	4-7	128-133	136-210	43
$RhH(CO)(PPh_3)_3$	—	—	25-50	1-80	11, 25

$Rh(CO)L_2X$ (L = PAr_3, PR_3, $P(OEt)_3$, $AsPh_3$ X = Cl, Br, I, CNS)	—	—	70-145	25-100	25, 37, 71
$Rh(CO) (PPh_3)_2 Cl$	NEt_3	280	25	100	25, 109
	PPh_3	0-190	100-150	35-70	17, 79
$Rh(CO) (PPh_3)_2 Cl$ on Al_2O_3 or carbon	—	—	148	36-49	78, 79, 86
$Rh(CO) (AsPh_3)_2 Cl$	$AsPh_3$	6	100	71	79
$Rh(CO) (AsPh_3)_2 Cl$ on carbon	—	—	148	49	86
$Rh(PPh_3)_3 Cl$	—	—	70-147	45-120	25, 37, 59
$Rh(PPh_3)_3 Cl_3$	—	—	55	90	74, 107
$Rh(PEt_3)_3 Cl_3$	—	—	150	27-40	37
$Rh(CO) (PPh_3)_2 Cl_3$	—	—	150	27-40	37
$Rh(CO) L_2(OOCR')$ (L = PR_3, PPh_3 R' = H, R, Ar, CF_3)	—	—	65-180	1-70	8, 9, 10, 62
$Rh(CO) L(AA')$ (L = PR_3, PPh_3, $P(OBu)_3$ $P(OPh)_3$, $AsPh_3$ HAA' = β-diketone)	—	—	70-95	14-44	63
$Rh(CO) (PBu_3) (AN)$ (HAN = β-diketone monoimine)	—	—	78-85	25-42	63

The inhibiting effect of HCl is also shown by the relatively low activity of rhodium complexes containing several chlorine atoms, such as Rh(CO) (PPh₃)₂Cl₃ [25], 1, 2, 3- Rh(PPh₃)₃Cl₃ and [RhCl(SnCl₃)₂]₂⁴⁻ [74, 107] and by the inactivity (at 70 °C and 100 atm) of Rh(PEt₂Ph)₃Cl₃ [25]. The use of ethanol as a solvent is useful in some of these cases [75, 107], obviously because of its ability to react with HCl or to form metal hydrides.

Little attention has been given to nitrogen containing ligands or additives for the modification of rhodium-based hydroformylation catalysts. According to patents [8, 10] the use of Rh(CO)₂L (OOCR') type complexes (L = aniline, *p*-toluidine, pyridine or α-picoline) leads to the predominant formation of aldehydes and no alcohols are formed in a consecutive hydrogenation reaction. Rh(CO)₂LCl complexes (L = aniline, toluidine or α-picoline) are claimed to show a similar selectivity for aldehyde formation [37]. N-bases such as pyridine and pyrrolidine derivatives and triethyl amine promote the formation of β-formylisobutyric acid methyl ester (I) at the expense of the α-formyl isomer (II) in the hydroformylation of methyl methacrylate [29, 31] (see Section 3.5).

It is of interest, that in the same system, phospines have an opposite effect, but no explanation has been given for these facts.

Pyperidine was found to somewhat enhance diethylketone formation from ethylene, carbon monoxide and hydrogen [19]. In contrast to results with cobalt carbonyls as catalyst, pyridine had no such effect.

Polyamines, such as diethylene triamine have been proposed as additives in hydroformylation with rhodium catalysts [6].

Probably the use of N-bases as ligands would deserve a systematic study in the case of rhodium catalysts. No ligands of this type can be used with cobalt catalysts, since these are deactivated by strong Lewis bases [65, 106] but the activity of rhodium carbonyls is not significantly lowered by even a 3000 fold excess of pyridine [41].

3.3. Hydroformylation of Olefins

Practical reaction rates with unsubstituted rhodium carbonyls as catalysts are obtained between 70 and 150 °C and 50-200 atm total pressure (H₂ : CO = = 1 : 1). Substantially lower temperatures and pressures (down to 25 °C and 1 atm) can be used with some phosphine complexes of rhodium. Catalyst concentration may range between 0.05-50 mM. Normal: iso aldehyde ratios are low with simple rhodium carbonyls, but can be significantly increased by adding phosphines as ligands. Isomerization and hydrogenation of the olefins is usually not significant. At higher temperatures (160-220°) or longer reaction times, alcohols are formed by hydrogenation of the aldehydes.

The reaction is usually carried out in solution, but there are a few reports corcerning gas phase reactions over supported rhodium catalysts [20, 21, 78, 86].

3.3.1. KINETICS

According to Heil and Markó [46], the hydroformylation of n-heptene-l with Rh₄(CO)₁₂ as catalyst obeys the kinetic equation:

$$\frac{d\,[\text{aldehyde}]}{dt} = k\,[\text{olefin}]\,[\text{Rh}]\,\frac{p_{H_2}}{p_{CO}}$$

This equation was found to be valid under the following reaction conditions:

temperature: 75 °C; p_{H_2} 33-126 atm; p_{CO} 40-168 atm

At carbon monoxide partial pressures below 40 atm the rate is proportional to p_{CO}, i.e. the apparent rate constants show a maximum at 40 atm p_{CO} (Fig. 1.).

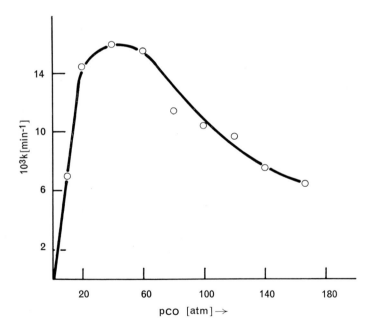

Figure 1. Influence of carbon monoxide partial pressure on the rate of hydroformylation of *n*-heptene-1 with Rh₄(CO)₁₂ as catalyst [46].

The energy of activation between 66-90 °C was determined as 27.8 kcal · mol⁻¹. The rate of reaction is strongly influenced by the structure of the olefin [48], styrene showing the highest reactivity and branched olefins the lowest (Table 6.)

Table 6

Effect of Olefin Structure on Rate [48]

Catalyst: $Rh_4(CO)_{12}$
Catalyst concentration: 0.21 mg atom $Rh \cdot 1^{-1}$
Temperature: 75 °C
Pressure: 130 atm ($1H_2 + 1CO$)

Olefin	Apparent rate constant, $10^3 \cdot k \cdot min^{-1}$
Styrene	124
A. Linear terminal olefins	
Hexene-1	55.8
Heptene-1	54.2
Octene-1	50.1
Decene-1	40.5
B. Linear internal olefins	
cis, trans-Hexene-2	34.4
cis-Heptene-2	40.2
cis, trans-Heptene-3	41.9
C. Branched olefins	
2-Methylpentene-1	25.7
2-Methylpentene-2	14.5
trans-4-Methylpentene-2	29.7
2,4,4-Trimethylpentene-2	3.0
2,3-Dimethylbutene-2	0.7
D. Cyclic olefins	
Cyclohexene	6.1
Cyclopentene	16.2
1-Ethylcyclohexene-1	2.9

A somewhat different kinetic equation has been found for the hydroformylation of n-octene-1 with "rhodium nitrate" (dissolved in water) as catalyst [115]:

$$r = k\,[Rh]\,\frac{p_{H_2}}{1 + p_{CO}}$$

The reaction conditions were 110 °C, 53-178 atm p_{H_2}, 57-226 atm p_{CO} and 0.025-0,10 mg atom $Rh \cdot 1^{-1}$. The energy of activation between 90-110 °C was found to be 21 kcal $\cdot$ mol^{-1}.

Rudkovskii and coworkers [40, 41, 42] investigated the influence of reaction parameters on hexene-1 hydroformylation at 150 ºC in the presence of "rhodium carbonyl" or rhodium on asbestos as catalyst, by determining the time necessary for 50% conversion of the olefin. The rate of reaction above 60 atm was found to be proportional to p_{H_2} and to the 0,4th power of catalyst concentration (between 4-40 mg atom Rh · l^{-1}), showing a maximum at a p_{CO} of 150 atm. The energy of activation has been determined between 110-165 ºC as 6.8 kcal · mol^{-1}. Hexene-1, isobutylene and cyclohexene reacted at practically the same rate. The reaction was fastest in dioxane and slowest in ethanol with a rate ratio of about 3 : 1.

Despite basic agreements, there are also many discrepancies between the above data, which may be partly caused by the somewhat different reaction conditions. Principally, the results obtained at lower temperatures may be more reliable for two reasons: a) at higher temperatures olefin isomerization is faster (Section 3.3.2) and, since the rates of hydroformylation of isomeric olefins are significantly different, this may influence the results of rate measurements; b) higher temperatures mean higher reaction rates and the rate of gas diffusion into the liquid reaction mixture may become a limiting factor.

Due to the lack of systematic work in this respect, no clear picture can be given regarding the influence of phospines or phosphites on reaction rate when added to rhodium catalysts not already containing a phosphorus ligand. Some published data suggest that higher temperatures are needed in the presence of tributylphosphine than when Rh_2O_3 alone is used as a catalyst [36], whereas the reverse was observed in the case of the Rh on carbon + $P(OPh_3)_3$ catalyst combination [45]. Still other experiments reveal apparently no significant influence of added phosphine [29].

No complete and detailed kinetic picture has yet been given for any of the phosphine containing catalysts. The most detailed information is available with $RhH(CO)(PPh_3)_3$ at 25 ºC and 1 atm. The reaction is inhibited by carbon monoxide or excess PPh_3 [25], and the rate is influenced by the structure of the olefin [11] (Table 7.) in a similar manner to that observed with $Rh_4(CO)_{12}$ under pressure [48] (Table 6.). With $Rh(CO)(PPh_3)_2Cl$ at elevated temperatures (70-150 ºC) and pressures (35-70 atm) the rate is first order with respect to the olefin [17, 25], increases with the H_2 : CO ratio [17, 50] and decreases with an increase of p_{CO} [25].

Moreover the catalytic activity of $Rh(CO)(PAr_3)_2X$ (X = Cl, Br, J) type complexes decreases in the order Cl > Br > J and Ar = p-$MeOC_6H_4$ > C_6H_5 > > p-FC_6H_4 [25].

3.3.2. ISOMERIZATION OF OLEFINS AND DISTRIBUTION OF ISOMERIC REACTION PRODUCTS

Starting from terminal olefins and using simple rhodium carbonyls as catalysts, the normal : iso aldehyde ratio is lower (1-1.2) than with cobalt carbonyls (1.5-4). In the case of propylene [81, 82] this may be seen from Tables 3 and 4 and similar results were obtained with higher n-olefins [36, 105]. No systematic study of the influence of reaction conditions on the normal : iso ratio has been performed yet, the very little information that exists leads to the conclusion that temperature has only a minor effect [42, 105].

Table 7

EFFECT OF OLEFIN STRUCTURE ON RATE [11]

Catalyst: 2.5 mM $RhH(CO)(PPh_3)_3$
Temperature: 25 °C
Pressure: 1 atm ($1H_2 + 1CO$)

Olefin	Rate of gas absorption ml · min^{-1}
Allyl alcohol	7.05
Allyl phenyl ether	5.78
Styrene	4.32
Hexadiene-1.5	4.26
4-Vinylcyclohexene	4.21
o-Allylphenol	4.03
Pentene-1	3.74
Allyl cyanide	3.72
Allyl benzene	3.56
Hexene-1	3.52
Heptene-1	3.50
Dodecene-1	3.18
Vinyl acetate	0.75
Cyclooctene	0.26
Ethyl vinyl ether	0.20
cis, trans-Pentene-2	0.15
cis-Heptene-2	0.12
d,l-Limonene	0.10
2-Methylpentene-1	0.06
Cyclohexene	0

If internal olefins are hydroformylated at 70-100 °C, practically only branched aldehydes are obtained [36, 105]. This is in accordance with the observation that, below 100 °C, olefin isomerization is very slow [36, 48, 105, 115]. The pronounced difference in product composition when starting from terminal or internal olefins is a characteristic feature of rhodium catalyzed hydroformylation. With cobalt catalysts the ratio of isomeric aldehydes depends only to a small extent on the position of the double bond [80]. This may find practical application when selective production of branched aldehydes or alcohols is desired.

The olefin isomerization reaction is strongly temperature dependent [36]. With "rhodium nitrate" as catalyst and under 100 atm of carbon monoxide the energy of activation between 90-110 ^{o}C was determined as 38 kcal $\cdot$ mol^{-1} [115]. On hydroformylating n-hexene-1 at 150 ^{o}C, the normal : iso aldehyde ratio decreases with increasing conversion [42].

Starting from branched chain olefins, aldehydes with quaternary carbon atoms are formed only in small amounts [42, 77]. Styrene gives predominantly the branched product, α-phenylpropionaldehyde [99].

With rhodium carbonyls, the addition of phosphines or phosphites to the reaction mixture increases the amount of normal aldehydes formed from terminal olefins, but the effect was claimed to be less significant than in the case of cobalt catalysts [93]. This observation, however, is not supported by numerous other experiments which yield much higher normal : iso ratios than have as yet been obtained with cobalt catalysts.

Table 8

EFFECT OF THE STRUCTURE OF TERTIARY PHOSPHORUS LIGANDS ON ALDEHYDE ISOMER DISTRIBUTION [85]

Olefin: n-octene-1
Catalyst: 5% Rh on carbon
Catalyst concentration: 14 mg atom Rh $\cdot$ l^{-1}
PR$_3$: Rh = 10 : 1
Pressure: 5.5-7 atm (1H$_2$ + 1CO)

Nature of R in PR$_3$	Temperature, ^{o}C	Reaction time, min	Straight chain aldehyde, %
n-Butyl	90	225	71
Phenyl	90	35	82
o,o-Dimethylphenoxy	90	80	47
o-Phenylphenoxy	90	95	52
o-Methylphenoxy	90	52	78
p-Methoxyphenoxy	90	270	83
p-Phenylphenoxy	90	70	85
Phenoxy	90	50	86
p-Chlorophenoxy	90	55	93
n-Butoxy	110	60	81

Phosphines are less effective than phosphites and the influence of the latter is further increased if they contain stronger electronegative groups. Attaching groups to the *ortho* positions of triphenylphosphite, thus increasing the steric

18

requirements of the ligand, was desadvantageous. These effects can be seen in Table 8 [84, 85].

As can be expected, products distribution is also sensitive to the P : Rh ratio. Adding more triphenylphosphite results in the formation of more straight chain product. This effect is partly offset by increasing the synthesis gas pressure, thus showing that CO and $P(OPh)_3$ are competing ligands in the catalytically active complexes and the predominant formation of straight chain isomers is due to the phosphite-containing species (Table 9.) [84, 85].

Table 9

EFFECT OF PRESSURE AND LIGAND CONCENTRATION ON ALDEHYDE ISOMER DISTRIBUTION
[85]

Olefin: *n*-octene-1
Catalyst: 5% Rh on carbon
Catalyst concentration: 14 mg atom Rh · 1^{-1}
Temperature: 90 °C

$P(OPh)_3$: Rh molar ratio	Pressure, atm ($1H_2 + 1CO$)	Reaction time, min	Straight chain aldehyde, %
0	5.5–7	180	31
2.2	5.5–7	30	74
6.7	5.5–7	50	86
13.5	5.5–7	35	87
27	5.5–7	65	89
6.7	19–20	20	80
6.7	38–40	25	74
6.7	170	25	69

Using internal olefins as starting material, the formation of straight chain aldehydes can practically be eliminated if a large amount of tributylphosphine is added to the Rh_2O_3 catalyst [36]. This is explained by an inhibiting effect of the phosphine on olefin isomerization.

Among the phospine complexes used as catalysts, $RhH(CO)(PPh_3)_3$ and $Rh(CO)(PPh_3)_2Cl$ are the best investigated ones with respect to products distribution and olefin isomerization.

With terminal olefins and $RhH(CO)(PPh_3)_3$ as a catalyst, the normal: iso aldehyde ratio is strongly influenced by the reaction conditions [11, 25]. At 25 °C a normal : iso ratio of about 20 is obtained if a stoichiometric reaction is carried out using $[Rh(CO)_2(PPh_2)_2]_2$ obtained by CO saturation of a benzene solution of RhH(CO)$_2$ (PPh$_3$)$_2$ [25] on the other hand in a catalytic reaction at 25 °C and 1 atm total pressure of a 1 : 1 mixture of CO and H$_2$, the normal : iso ratio is around 3 : 1. In this case the ratio is increased by increasing: a) the catalyst concentration, b) temperature, c) H$_2$: CO ratio, and d) by adding excess triphenylphosphine to the reaction mixutre. The highest value obtained was around 25 : 1 [11]. Starting from internal olefins, only branched aldehydes were formed [25]; RhH(CO) (PPh$_3$)$_3$ catalyses the isomerization of olefins [25, 114], but this is apparently inhibited by CO.

With Rh(CO) (PPh$_3$)$_2$Cl as catalyst, the normal : iso aldehyde ratio obtained from n-hexene-1 can be increased by applying higher pressures, adding excess triphenylphosphine or using polar solvents [17]. The somewhat unexpected effect of pressure is further confirmed by independent results [25]. Perhaps the most interesting is the effect of solvent polarity which suggests that solvent molecules with some Lewis basicity may, at least partly, take over the rôle of the triphenylphosphine ligands in forming catalytically active species. It is also of interest that the rate of reaction is increased about six fold by changing the solvent from benzene to butyraldehyde, which can hardly be explained simply by the difference in polarity. The possible rôle of solvent molecules as ligands is also supported by a patent claim in which Rh(Me$_2$SO)$_2$(CO)Cl is used as catalyst [37].

Starting from internal olefins, the normal : iso aldehyde ratio increases with temperature. At 70 °C no straight chain product was observed [25], the amount of which increased to 36.8% at 150 °C and 38.5% at 200 °C [17]. Excess triphenylphosphine reduced the amount of the normal isomer to 16.1% at 150 °C, but had no effect at 200 °C [17]. These observations can again be explained by the concurrent isomerization of the olefin which is promoted by hydrogen [97] but is retarded by phosphines [97], or carbon monoxide [25]. No experimental data are available, however, on the composition of the unconverted olefin at higher temperatures. As expected, the increase of temperature has an inverse effect on the normal : iso ratio in the case of terminal olefins [17, 50].

Interestingly there is no significant difference in the butyraldehyde isomer distribution, whether the Rh(CO) (PPh$_3$)$_2$Cl catalyst is applied in solution [50] or as a heterogeneous catalyst on Al$_2$O$_3$ support [86].

3.3.3. HYDROGENATION OF OLEFINS

Many of the rhodium complexes used as hydroformylation catalysts are in the absence of carbon monoxide, highly active catalysts for the hydrogenation of olefins, e. g. Rh(PPh$_3$)$_3$Cl [76], Rh(CO) (PPh$_3$)$_2$Cl [96] and RhH(CO) (PPh$_3$)$_3$ [73]. It is not the purpose of this review to discuss these hydrogenations and attention will be restricted to the hydrogenation of olefins as an undesirable side reaction in hydroformylation processes.

It is a general characteristic of hydroformylation processes with rhodium catalysts that hydrogenation is not significant at low temperatures in the case of simple monoolefins. This is true for catalysts without [54] and with [8, 9, 37, 50, 62, 63, 78] phosphorus containing ligands. Hydrogenation is substantially increased by applying higher temperatures. Thus, for example, with tetra-hydrobenzaldehyde as a substrate, 1-6% of hydrogenated product is formed at 120-130 °C and this increases to 35-41% if the temperature is raised to 200 °C [32].

Compounds with the carbon-carbon double bond conjugated to a carbonyl group are more susceptible to hydrogenation. The ethylesters of crotonic- and cinnamic acid are transformed in 19 and 26% yields respectively to the corresponding saturated esters, even at 120 °C [29]. These apparently discouraging results are still very good if compared with experiments using $Co_2(CO)_8$ as catalyst, with which 49-91.5% of the cinnamic ester is hydrogenated instead of being hydroformylated [29].

3.4. Hydroformylation of Diolefins

Diolefins with unconjugated double bonds can be hydroformylated to a mixture of isomeric dialdehydes [28, 30, 72]. In the case of cyclooctadiene-1, 5 the formation of the partially hydrogenated product, hydroxymethyl cyclooctane, could be suppressed by applying higher pressures [28]. Using the appropriate conditions, it appears to be possible to stop the reaction at the stage of the mono-hydroformylated product (i.e. the unsaturated aldehyde) [52]. Table 10 contains the experimental results obtained up till now.

Conjugated diolefins do not give dialdehydes if hydroformylated with simple rhodium carbonyls. Instead the formation of only monoaldehydes [72] and monoolefins [41] has been observed. However using a $Rh_2O_3 + PBu_3$ catalyst (P : Rh molar ratio = 85), butadiene and pentadiene-1.3 could be converted to the corresponding C_6 and C_7 dialdehydes in about 35-40% yield [35]. The other main products were the C_5 and C_6 monoaldehydes. The exact composition of the aldehyde fractions has not been determined.

Sorbic acid methyl ester gave no dialdehydes, only monoaldehydes being obtained [35]. Evidently, the ester group in a conjugated position promotes the hydrogenation of one of the olefinic double bonds, in accordance with the results [29] discussed in Section 3.3.3.

Hexadiene-1,5, 4-vinylcyclohexene and d,l-limonene could be hydroformylated with RhH(CO) (PPh₃)₃ as catalyst at 25 °C and 1 atm [11]. The structures of the products have not, however, been reported.

3.5. Hydroformylation of Substituted Olefins

Unsaturated alcohols, phenols, ethers, aldehydes, carboxylic acid esters, N-substituted amides and imides, carboxylic acids, anhydrides and nitriles were successfully hydroformylated. Table 11 gives a survey of these compounds and the reaction conditions used.

Table 10

HYDROFORMYLATION OF UNCONJUGATED DIENES

Diene	Catalyst	Temperature °C	Pressure, atm	Products	Yield, %	Reference
$CH_2=CHCH_2CH=CH_2$	Rh_2O_3	60-100	150	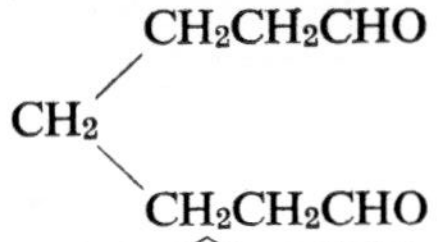 CH_2CH_2CHO $\mid$ CH_2	—	[72]
	"rhodium chloride"	?	?	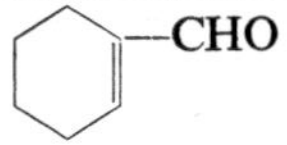 CH_2CH_2CHO CHO (cyclohexene)	—	[72]
	Rh_2O_3	60-100	150	$CH_2CH_2CH_2CHO$ CH_2CH_2CHCHO $\mid$ CH_3	—	[72]
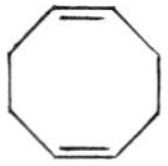 (cyclooctadiene)	Rh_2O_3	100/210 100/210 100/210 95/210 90/210	230/300 620 1050 850/910 1000/2100	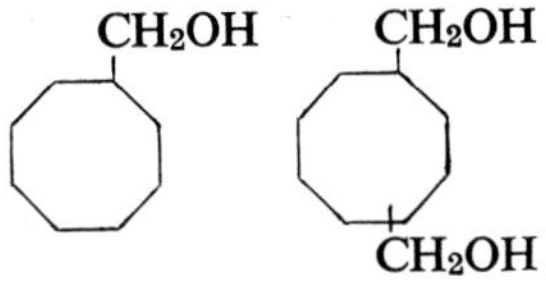 CH_2OH (cyclooctane) CH_2OH CH_2OH (cyclooctane)	62 16 25 59 11 63 15 70 13 81	[28] (*)
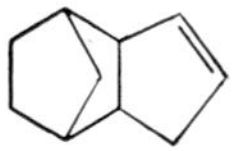 (tricyclic structure)		115	200	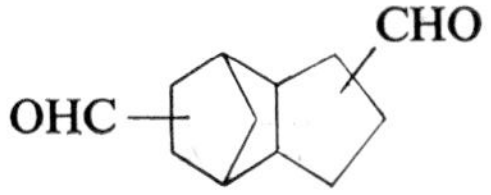 OHC— (tricyclic) CHO	61	[30]
(tricyclic structure)	Rh_2O_3	100	185	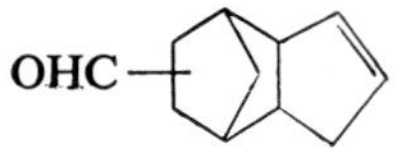 OHC— (tricyclic)	68	[52]

(*) The aldehydes obtained were hydrogenated to alcohols by raising temperature and pressure (Section 7).

22

Table 11

REACTION CONDITIONS AND PRODUCTS OF THE HYDROFORMYLATION OF SUBSTITUTED OLEFINS WITH RHODIUM CATALYSTS

Substrate	Catalyst	Temperature, °C	Pressure, atm	Products	Yield, %	Reference
$CH_2=CHCH_2OH$	$RhH(CO)(PPh_3)_3$	25	1	not stated	—	[11]
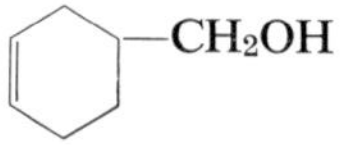 ⬡—CH_2OH	Rh_2O_3	130/200	200/300	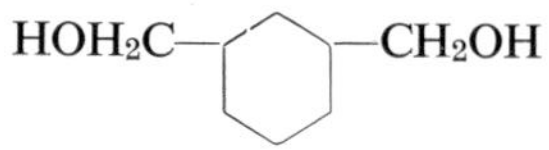 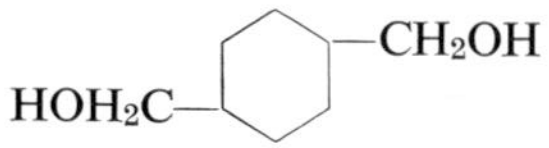 HOH_2C—⬡—CH_2OH	40	[32]¹)
				HOH_2C—⬡—CH_2OH	32-36	
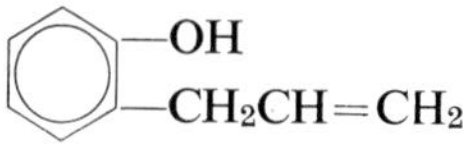 ⬡(OH)—$CH_2CH=CH_2$	$RhH(CO)(PPh_3)_3$	25	1	not stated	—	[11]
$CH_2=CHOC_2H_5$	$RhH(CO)(PPh_3)_3$	25	1	not stated	—	[11]
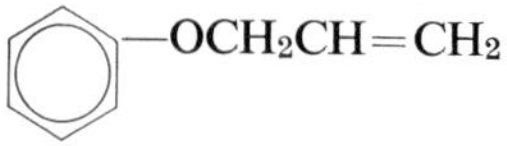 ⬡—$OCH_2CH=CH_2$	$RhH(CO)(PPh_3)_3$	25	1	not stated	—	[11]
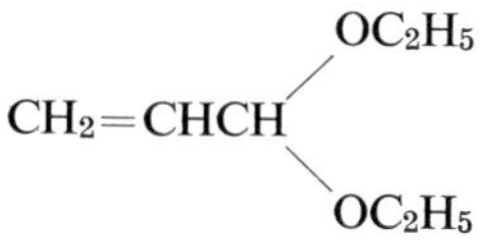 $CH_2=CHCH(OC_2H_5)(OC_2H_5)$	Rh_2O_3	110	200 (at 20 °C)	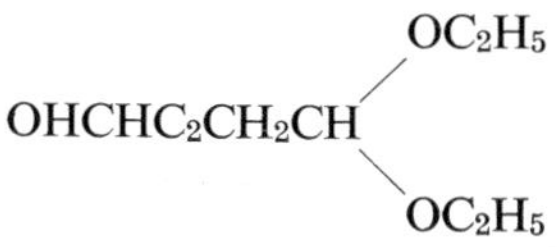 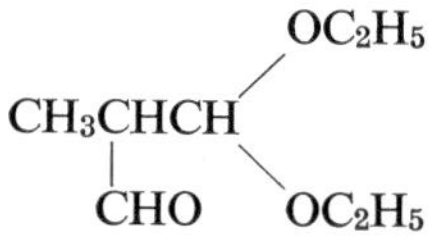 $OHCHCH_2CH_2CH(OC_2H_5)(OC_2H_5)$	22	[66]
				$CH_3CHCH(OC_2H_5)(CHO)(OC_2H_5)$	40	

Substrate	Catalyst	Temperature, °C	Pressure, atm	Products	Yield, %	Reference
 ⬡ bicyclic —CHO	Rh_2O_3	130/240	200/300	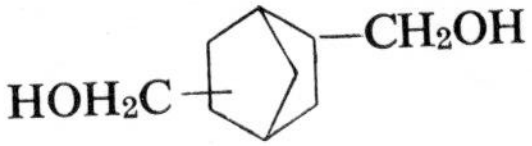 HOH_2C— ⬡ —CH_2OH	63	[30] [1]
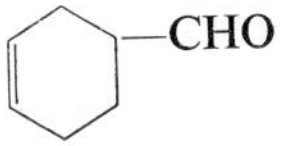 ⬡ —CHO	Rh_2O_3	120/200	200/300	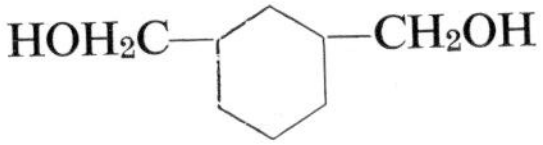 HOH_2C— ⬡ —CH_2OH	44-47	[32] [1]
				⬡ —CH_2OH	36-38	
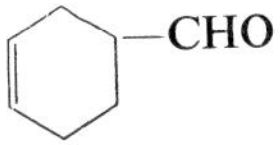 ⬡ —CHO	Rh_2O_3	100	240	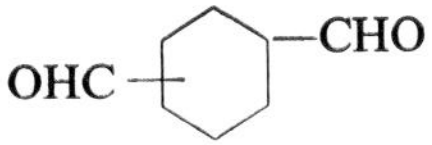 OHC— ⬡ —CHO	71	[4]
				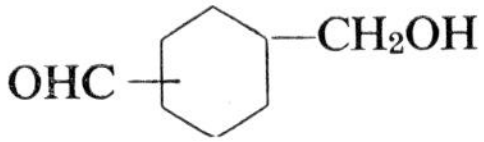 OHC— ⬡ —CH_2OH	14	
$CH_3COOCH=CH_2$	$RhH(CO)(PPh_3)_3$	25	1	not stated	—	[11]
	$Rh_4(CO)_{12}$	80	165	$CH_3COOCHCH_3$ / CHO	72	[2]
CH_3COO—CHCH=CH_2 / CH_3COO	Rh_2O_3	100	200 (at 20 °C)	CH_3COO—$CHCH_2CH_2CHO$ / CH_3COO CH_3COO—CHCHCH$_3$—CHO / CH_3COO	—	[66]

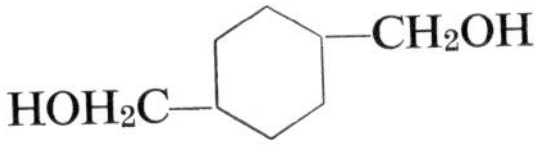

Continued

Table 11 REACTION CONDITIONS AND PRODUCTS OF THE HYDROFORMYLATION OF SUBSTITUTED OLEFINS WITH RHODIUM CATALYSTS

Substrate	Catalyst	Temperature, °C	Pressure, atm	Products	Yield, %	Reference
$CH_2{=}CHCOOCH_3$	Rh_2O_3	120	200	"formylpropionic acid methyl ester"	62	[18]
$CH_2{=}CHCOOC_2H_5$	"rhodium carbonyl"	95-120	120-240	$OHCCH_2CH_2COOC_2H_5$ $CH_3\overset{\mid}{C}HCOOC_2H_5$ CHO	16-71 13-57	[98]
	$Rh_4(CO)_{12}$	80	—	$CH_3\overset{\mid}{C}HCOOC_2H_5$ CHO	60	[100]
$\overset{CH_3}{\underset{}{CH_2{=}CCOOCH_3}}$	Rh_2O_3 (+ PZ_3 or N-bases)	80-150	200-1000	$OHCCH_2\overset{CH_3}{\overset{\mid}{C}H}COOCH_3$ $CH_3\overset{\mid}{C}HCOOCH_3$ CHO	0-75	[29] ²⁾
	Rh_2O_3	70 or 130		$OHCCH_2\overset{CH_3}{\overset{\mid}{C}H}COOCH_3$ $CH_3\overset{\mid}{C}HCOOCH_3$ CHO	17 or 68 74 or 17	[18]
	Rh_2O_3 (+N-bases)	130-160/190-230	200/300	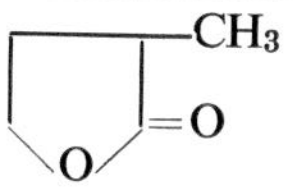 $\overset{CH_3}{\underset{O=}{\square}}$	38-80	[31] ¹⁾
	Rh_2O_3 (+N-bases)	160	200	$OHCCH_2\overset{CH_3}{\overset{\mid}{C}H}COOCH_3$	61-79	[31]

Substrate	Catalyst	Temperature, °C	Pressure, atm	Products	Yield, %	Reference
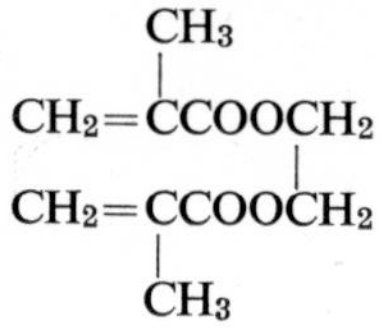 CH_3 $CH_2{=}CCOOCH_3$ $CH_2{=}CCOOCH_2$ CH_3	Rh_2O_3	120	250	"ethyleneglycolbisformylisobutyric ester"	75	[18]
$CH_3CH{=}CHCOOC_2H_5$	Rh_2O_3	70-180/ 200-230	200/300	(lactone structures) $CH_3CH_2CHCOOC_2H_5$ CH_2OH	1-30 15-60 0-21	[29] [1,3]
$CH_3CH_2CH{=}CHCH_2COOCH_3$ and $CH_2{=}CHCH_2CH_2CH_2COOCH_3$	Rh_2O_3 $+ PBu_3$	120-160	200 (at 20 °C)	Mixture of formylcapronic acid methyl esters	—	[32] [4]
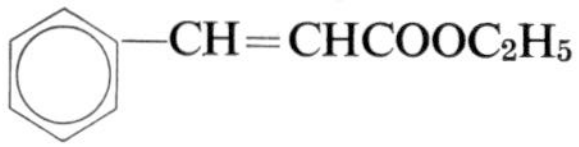 Ph$-CH{=}CHCOOC_2H_5$	Rh_2O_3	120/230	200/300	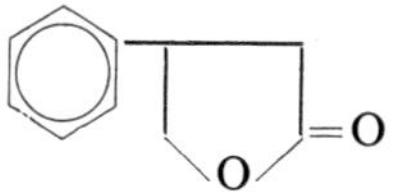 (lactone structure)	73	[29] [1]
$CH_3CH{=}CHCH{=}CHCOOCH_3$	Rh_2O_3	125	200 (at 20 °C)	"C_7 aldehydes"	—	[35]
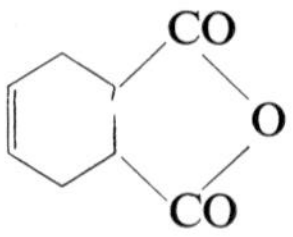 (bicyclic anhydride CO–O–CO structure)	Rh_2O_3	100	250	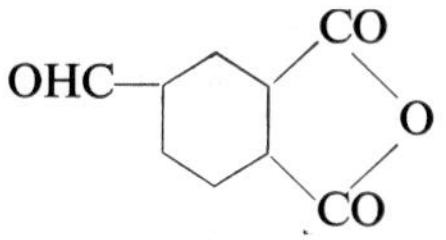 OHC (bicyclic anhydride structure)	74	[18]

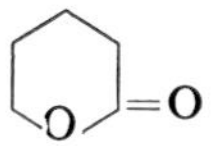

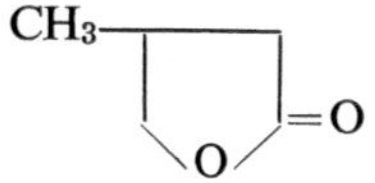

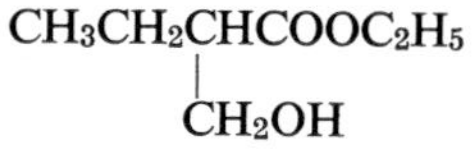

Continued

Table 11 REACTION CONDITIONS AND PRODUCTS OF THE HYDROFORMYLATION OF SUBSTITUTED OLEFINS WITH RHODIUM CATALYSTS

Substrate	Catalyst	Temperature, °C	Pressure, atm	Products	Yield, %	Reference
$CH_3CONHCH_2CH=CH_2$	$Rh_4(CO)_{12}$	90–100	170	$CH_3CONHCH_2CH_2CH_2CHO$ $CH_3CONHCH_2CHCH_3$ with CHO	9 78	[1,87]
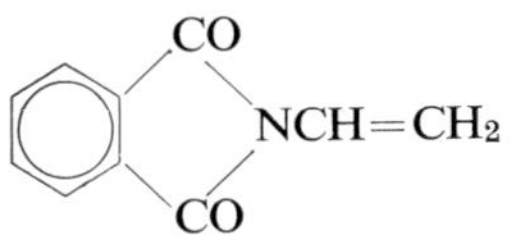 phthalimido–$NCH=CH_2$	$Rh_4(CO)_{12}$	100	170	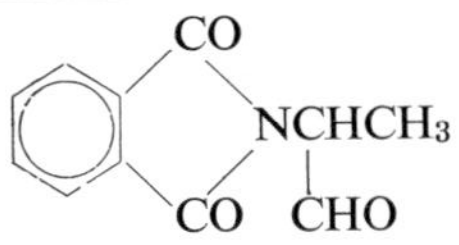 phthalimido–$NCHCH_3$ with CHO	81	[87]
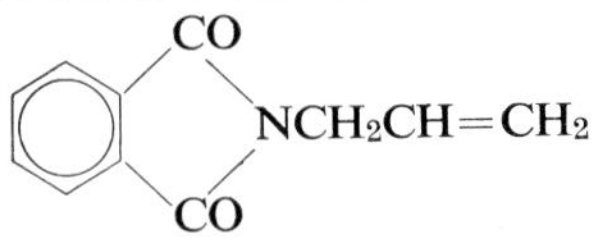 phthalimido–$NCH_2CH=CH_2$	$Rh_4(CO)_{12}$	65–70	170	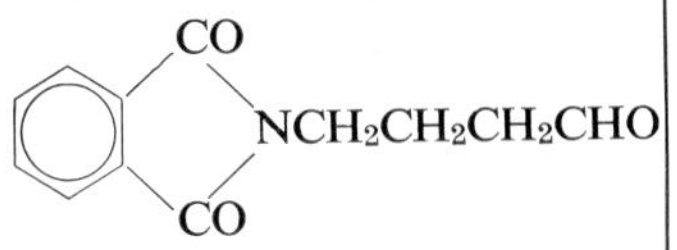 phthalimido–$NCH_2CH_2CH_2CHO$ phthalimido–NCH_2CHCH_3 with CHO	15 80	[1,87]
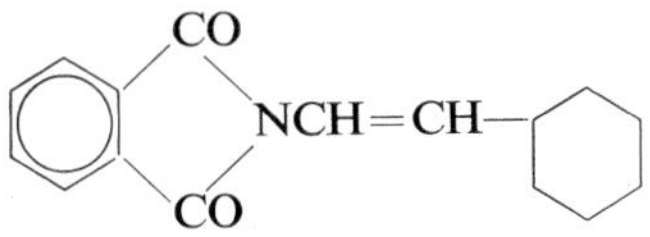 phthalimido–$NCH=CH$–cyclohexyl	$Rh_4(CO)_{12}$	120	170	"mixture of aldehydes"	74	[87]
$CH_2=CHCH_2CN$	$RhH(CO)(PPh_3)_3$	25	1	not stated	—	[11]

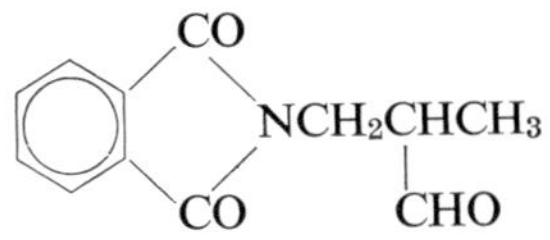

(¹) The aldehydes obtained were hydrogenated to alcohols by raising temperature and pressure.   (²) Tables 12-15.   (³) Table 16.   (⁴) Table 17.

The hydroformylation of methacrylic acid methyl ester has been investigated in gratest detail. The ratio of the two isomeric products, α-and β-formylisobutyric acid methylesters, depends strongly on the reaction conditions [29]:

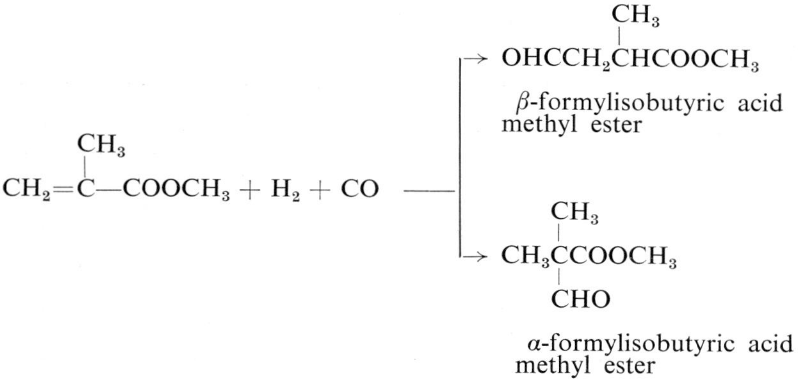

$$\text{CH}_2=\overset{\overset{\displaystyle \text{CH}_3}{|}}{\text{C}}-\text{COOCH}_3 + \text{H}_2 + \text{CO} \longrightarrow \begin{cases} \rightarrow \text{OHCCH}_2\overset{\overset{\displaystyle \text{CH}_3}{|}}{\text{C}}\text{HCOOCH}_3 \\ \text{β-formylisobutyric acid methyl ester} \\ \\ \rightarrow \text{CH}_3\overset{\overset{\displaystyle \text{CH}_3}{|}}{\underset{\underset{\displaystyle \text{CHO}}{|}}{\text{C}}}\text{COOCH}_3 \\ \text{α-formylisobutyric acid methyl ester} \end{cases}$$

The formation of the α-formylated product is favoured by low temperatures and high pressures (Tables 12 and 13). Added ligands have also a pronounced effect in the sense that phosphines and phosphites enhance the formation of the α-formyl ester and N-bases that of the β-formyl ester (Table 14). The effect of pressure is also noticeable in those cases when tributylphosphine is used to promote α-formylation (Table 15). No theoretical explanation has been given for these facts, the effect of phosphines in this case should, however, be useful in interpreting the influence of these ligands on the formation of isomeric products discussed in Section 3.7.2. The hydroformylation of methacrylic acid methyl ester to produce α-methyl γ-butyrolactone will be discussed in Section 3.7.

Table 12

EFFECT OF TEMPERATURE ON THE HYDROFORMYLATION OF METHACRYLIC ACID METHYL ESTER [29]

Catalyst: 0.04 mole % Rh_2O_3
Pressure: 1000 atm ($1H_2 + 1CO$)

Reaction products	80 °C	100 °C	120 °C	150 °C
β-Formylisobutyric acid methyl ester	13%	30%	50%	71%
α-Formylisobutyric acid methyl ester	72%	54%	35%	11%

Table 13

EFFECT OF PRESSURE ON THE HYDROFORMYLATION OF METHACRYLIC ACID METHYL ESTER [29]

Catalyst: 0.04 mole % Rh_2O_3
Temperature: 80 °C

Reaction products	200 atm	400 atm	500 atm	1000 atm
β-Formylisobutyric acid methyl ester	42%	34%	31%	13%
α-Formylisobutyric acid methyl ester	42%	48%	52%	72%

Table 14

EFFECT OF PHOSPHORUS- AND NITROGEN-CONTAINING LIGANDS ON THE HYDROFORMYLATION OF METHACRYLIC ACID METHYL ESTER [29]

Catalyst: 0.04 mole % Rh_2O_3
Pressure: 200 atm ($1H_2 + 1CO$)

Added ligand	Ligand: Rh ratio	Tempera-ture, °C	Yield, %	
			β-Formylisobutyric acid methyl ester	α-Formylisobutyric acid methyl ester
—	—	130	63	17
N-Ethylpyrrolidine	104	130–150	75	4
Pyridine	75	130	59	8
Tributylphosphine	31	130	10	65
1-Phenyl-phospholine-2	19	80–130	—	86
Triphenylphosphine	24	80	9	73
Triethylphosphite	38	120–150	—	58

Qualitatively similar effects of temperature and pressure on isomer distribution were noted in the hydroformylation of acrylic acid ethyl ester [98]: the

Table 15

EFFECT OF PRESSURE ON THE HYDROFORMYLATION OF METHACRYLIC ACID METHYL
ESTER IN THE PRESENCE OF TRIBUTYLPHOSPINE [29]

Catalyst: 0.04 mole % Rh_2O_3
Additive: 2.5 mole % PBu_3
Temperature: 120-130 °C

Reaction products	200 atm	250 atm	300 atm	400 atm	1000 atm
β-Formylisobutyric acid methyl ester	10%	9%	8%	8%	4%
α-Formylisobutyric acid methyl ester	65%	74%	80%	91%	92%

Table 16

EFFECT OF TEMPERATURE ON PRODUCT DISTRIBUTION IN THE HYDROFORMYLATION AND
SUBSEQUENT HYDROGENATION OF CROTONIC ACID ETHYL ESTER [29]

Catalyst: 1 weight % Rh_2O_3
Pressure: a) hydroformylation (first step) 200 atm ($1H_2 + 1CO$)
　　　　　b) hydrogenation (second step) 300 atm ($1H_2 + 1CO$)
Temperature of hydrogenation (second step) 230 °C

Temperature of hydrofor-mylation, °C	Yield, %				
	Butyric acid ethyl ester	α-Hydroxy-methylbu-tyric acid ethyl ester	β-Methyl-butyrolac-tone	δ-Valero-lactone	Total
70–90	18	21	15	1	55
100–110	18	9	32	1	60
120	19	6	46	3	74
125	19	5	52	5	81
130	17	4	57	8	82
135	20	3	60	9	92
140	20	2	55	11	88
150	20	1	43	15	79
180	30	—	31	13	91

formation of the straight chain product was favoured by using higher temperatures and lower pressures.

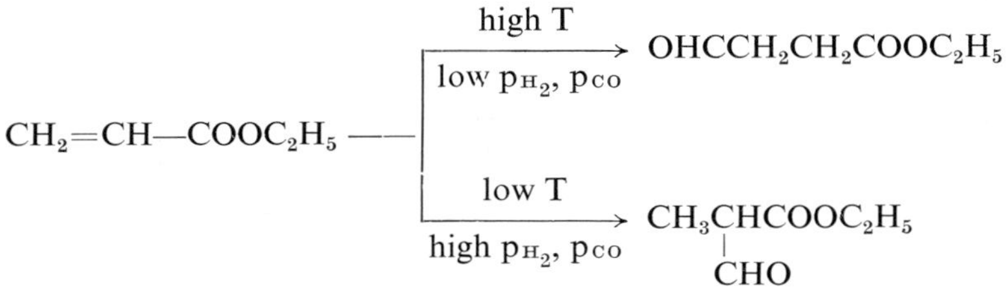

$$CH_2{=}CH{-}COOC_2H_5 \quad
\begin{array}{l}
\xrightarrow[\text{low } p_{H_2},\, p_{CO}]{\text{high T}} \quad OHCCH_2CH_2COOC_2H_5 \\[2em]
\xrightarrow[\text{high } p_{H_2},\, p_{CO}]{\text{low T}} \quad CH_3CHCOOC_2H_5 \\
\hspace{8em} | \\
\hspace{8em} CHO
\end{array}$$

The influences of P_{H2} and P_{CO} were studied separately in this case, the results showing that both had the same effect.

In addition the hydroformylation of crotonic acid ethyl ester at different temperatures gave the result that the increase of temperature enhances the formylation of the carbon atom more remote from the ester group [29]. This is shown by the data in Table 16. This latter also shows that raising the temperature favours the isomerization of the double bond as indicated by the formylation of the γ carbon atom, which was not originally involved in the $C = C$ bond:

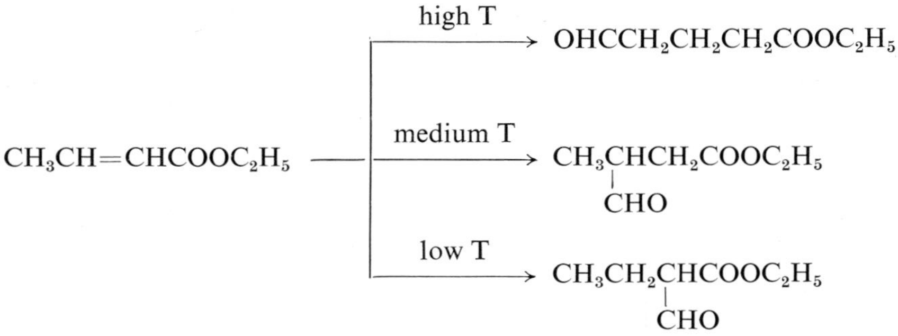

$$CH_3CH{=}CHCOOC_2H_5 \quad
\begin{array}{l}
\xrightarrow{\text{high T}} \quad OHCCH_2CH_2CH_2COOC_2H_5 \\[2em]
\xrightarrow{\text{medium T}} \quad CH_3CHCH_2COOC_2H_5 \\
\hspace{9em} | \\
\hspace{9em} CHO \\[1.5em]
\xrightarrow{\text{low T}} \quad CH_3CH_2CHCOOC_2H_5 \\
\hspace{8em} | \\
\hspace{8em} CHO
\end{array}$$

The products listed in Table 16 are not actually the aldehydes shown above but the products derived from them by subsequent hydrogenation and partial lactonization. These consecutive reactions will be discussed in Section 7.

The migration of the double bond can be suppressed also in the case of substituted olefins by adding phosphines to the reaction mixture (Section 3.3.2.). The results of experiments with $\Delta3$ and $\Delta5$ hexenoic acid methyl esters, have been collected together in Table 17 [36]. The reaction mixture is the following in the case of $\Delta3$-hexenoic acid methyl ester:

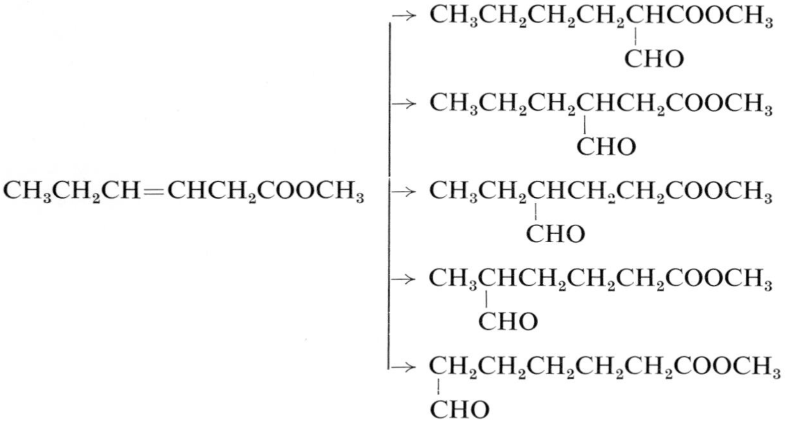

$$\rightarrow CH_3CH_2CH_2CH_2CHCOOCH_3 \qquad \text{1-formyl}$$
$$\qquad\qquad\qquad\quad |$$
$$\qquad\qquad\qquad\quad CHO$$

$$\rightarrow CH_3CH_2CH_2CHCH_2COOCH_3 \qquad \text{2-formyl}$$
$$\qquad\qquad\qquad\quad |$$
$$\qquad\qquad\qquad\quad CHO$$

$$CH_3CH_2CH=CHCH_2COOCH_3 \quad \rightarrow CH_3CH_2CHCH_2CH_2COOCH_3 \qquad \text{3-formyl}$$
$$\qquad\qquad\qquad\qquad\qquad\qquad\qquad\quad |$$
$$\qquad\qquad\qquad\qquad\qquad\qquad\qquad\quad CHO$$

$$\rightarrow CH_3CHCH_2CH_2CH_2COOCH_3 \qquad \text{4-formyl}$$
$$\qquad\quad |$$
$$\qquad\quad CHO$$

$$\rightarrow CH_2CH_2CH_2CH_2CH_2COOCH_3 \qquad \text{5-formyl}$$
$$\quad |$$
$$\quad CHO$$

Table 17

EFFECT OF TRIBUTYLPHOSPHINE ON ISOMER DISTRIBUTION IN THE HYDROFORMYLATION OF HEXENOIC ACID METHYL ESTERS [36]

Catalyst: 0.24 mole % Rh_2O_3
PBu_3 : Rh = 30 : 1
Pressure: 200 atm ($1H_2 + 1CO$) at 20°

Substrate	Temperature, °C	PBu_3 added	Isomer distribution, %				
			1-formyl	2-formyl	3-formyl	4-formyl	5-formyl
$CH_3CH_2CH=CHCH_2COOCH_3$	120	no	1	36	28	28	8
	160	yes	—	66	34	—	—
$CH_2=CHCH_2CH_2CH_2COOCH_3$	120	no	trace	3	12	41	44
	160	yes	—	—	—	39	61

3.6. Hydroformylation of Acetylenes

There is only one report on the hydroformylation of acetylenes [59, 107]. With $Rh(PPh_3)_3Cl$ in an ethanol-benzene solution, hexyne-1 reacted at 180°C and 120 atm ($1H_2 + 4CO$) to give about 15% yield of n-heptaldehyde and 2-methylhexaldehyde in equal amounts. The reaction time was 16 hours showing that it is much more difficult to hydroformylate acetylenes than olefins.

3.7. Mechanism of the Reaction

There are certainly many details which still remain to be clarified, but the general features of the mechanism of hydroformylation with rhodium complexes as catalyst are now fairly well elucidated and generally accepted. This mechanism follows the scheme first elaborated by Breslow and Heck [7] for cobalt carbonyl catalysts and which can be represented schematically by the following reaction cycle:

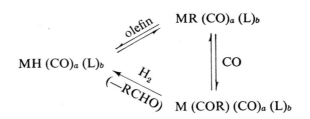

Most experimental results are in agreement with the above scheme. In a recent work [115] based on kinetic data, the role of dinuclear catalytic intermediates of the type $Rh_2(CO)_7$ (olefin) and $Rh_2H_2(CO)_6$ (olefin) is discussed.

The main reason for the higher activity of rhodium catalysts as compared to cobalt is probably the greater ease of oxidative addition of hydrogen to the acyl complex $M(COR)(CO)_a(L)_b$ in the case of rhodium [25]. This seems to be more likely than the suggestion that the larger size of the central metal atom reduces steric hindrances. If this were true iridium should be even more active, which it is not (Section 4).

3.7.1. Hydroformylation with Unsubstituted Rhodium Carbonyls as Catalysts

Both the kinetics [46] and the influence of olefin structure on the rate of hydroformylation [48] stress the close similarity between the cobalt carbonyl- and rhodium carbonyl-catalyzed reactions. The main difficulty in the interpretation of this similarity is the fact that, since the first report of their preparations in 1943 [51], neither $Rh_2(CO)_8$ nor $RhH(CO)_4$ could be characterized again [13, 45]. The presence of the analogous cobalt complexes $Co_2(CO)_8$ and $CoH(CO)_4$ in the reaction mixture under conditions of hydroformylation is well documented [69] but no similar evidence could be obtained for the corresponding rhodium carbonyls [46, 48].

Recent experiments partly resolved this contradiction by proving the presence of $[Rh(CO)_2(OOCR)]_2$ complexes [64] in liquid samples obtained from the hydroformylation reaction mixture [49]. These carboxylato complexes were found in the samples as long as the reaction was proceeding (i.e. unreacted olefin was present). Their concentration was observed to be proportional to the rate of reaction (i.e. was influenced by the reaction conditions like the reaction rate itself). Although these features are characteristic of a reaction intermediate, the carboxylato complexes themselves cannot be regarded as intermediates, since they were shown to be transformed under the reaction conditions to $Rh_4(CO)_{12}$ and carboxylic acids. Their presence is explained by the action of traces of oxygen, which transform the presumably extremely labile, acylrhodium carbonyls to these more stable derivatives. Thus the presence of $[Rh(CO)_2(OOCR)]_2$ complexes can be regarded as indirect evidence for acyl rhodium carbonyls being intermediates of hydroformylation:

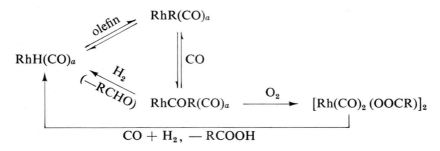

Since rhodium carbonyls generally have a greater tendency to lose carbon monoxide than the corresponding cobalt carbonyls, the number of CO ligands coordinated to rhodium in the above complexes is probably less than four. Kinetic data support an equilibrium:

$$RhH(CO)_3 + CO \rightleftharpoons RhH(CO)_4$$

which is not shifted strongly to either side at 160 °C and 80-160 atmospheres p_{CO} [45]. It was suggested [105] that the presence of a tricarbonyl instead of a tetracarbonyl could explain the greater amount of branched aldehydes formed from terminal olefins in the case of rhodium as compared to cobalt. This would be in agreement with the influence of p_{CO} on isomer distribution in the case of cobalt carbonyls as catalysts [80, 83].

The influence of olefin structure on rate [48] is clearly mainly dominated by electronic factors: electron releasing alkyl groups decrease the reaction rate and the electron withdrawing phenyl group increases it. No mechanistic interpretation has been given for this fact yet.

3.7.2. HYDROFORMYLATION WITH $RhH(CO)(PPh_3)_3$ AS CATALYST

$RhH(CO)(PPh_3)_3$ is the phosphine containing catalyst which has been studied in most detail. However, before discussing the mechanism of the hydroformylation reaction with this catalyst, a few words must be said about the chemistry of this complex under hydroformylation conditions.

If $RhH(CO)(PPh_3)_3$, in the presence of hydrogen and carbon monoxide,

34

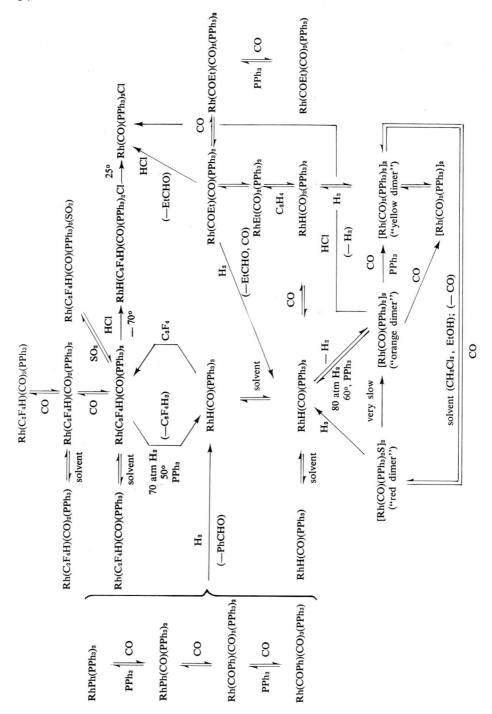

Figure 2. Reaction of RhH(CO) (PPh₃)₃ [26, 110, 111, 113].

is dissolved in benzene, dichloromethane or ethanol, a complicated set of dissociation and dimerization reactions takes place [26, 113]. These reactions and some others [110, 111] which bear a direct relationship to the activity of the complex as a hydroformylation catalyst, are illustrated in Fig. 2.

The molecular weight of $RhH(CO)(PPh_3)_3$ measured in benzene is concentration dependent, which suggests the equilibria [26]:

$$RhH(CO)(PPh_3)_3 \underset{+ PPh_3}{\overset{- PPh_3}{\rightleftharpoons}} RhH(CO)(PPh_3)_2 \underset{+ PPh_3}{\overset{- PPh_3}{\rightleftharpoons}} RhH(CO)(PPh_3)$$

If this solution is treated with carbon monoxide at 25 °C and 1 atm, further equilibrium reactions take place [26]:

$$RhH(CO)(PPh_3)_2 + CO \rightleftharpoons RhH(CO_2)(PPh_3)_2$$

$$2\ RhH(CO)_2(PPh_3)_2 \underset{H_2}{\overset{CO}{\rightleftharpoons}} [Rh(CO)_2(PPh_3)_2]_2 + H_2$$

« yellow dimer »

At somewhat elevated temperature (80 °C) and pressure (7 atm, $1H_2 + 1CO$) the exchange of PPh_3 and CO ligands seems to go even further, as inferred from infrared spectroscopic measurements [85]:

$$RhH(CO)(PPh_3)_3 \underset{PPh_3}{\overset{CO}{\rightleftharpoons}} RhH(CO)_2(PPh_3)_2 \underset{PPh_3}{\overset{CO}{\rightleftharpoons}} RhH(CO)_3(PPh_3)$$

The complex $RhH(CO)_2(PPh_3)_2$ is extremely air sensitive and could not be isolated in a crystalline form. The "yellow dimer" was proposed to have a structure with both terminal and bridging carbonyl ligands [26]:

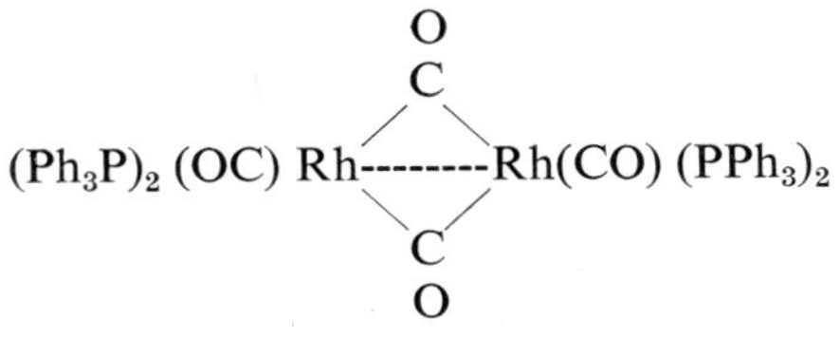

$$
\begin{array}{c}
O \\
C \\
\diagup \diagdown \\
(Ph_3P)_2(OC)Rh\text{--------}Rh(CO)(PPh_3)_2 \\
\diagdown \diagup \\
C \\
O
\end{array}
$$

Under nitrogen or argon, or if evacuated, solutions containing the "yellow dimer" loose carbon monoxide and turn red. In the presence of dichloromethane or ethanol, solvates of the type $[Rh(CO)(PPh_3)_2S]_2$ can be isolated (S = CH_2Cl_2, EtOH), which are called the "red dimers". These dimers have only bridging

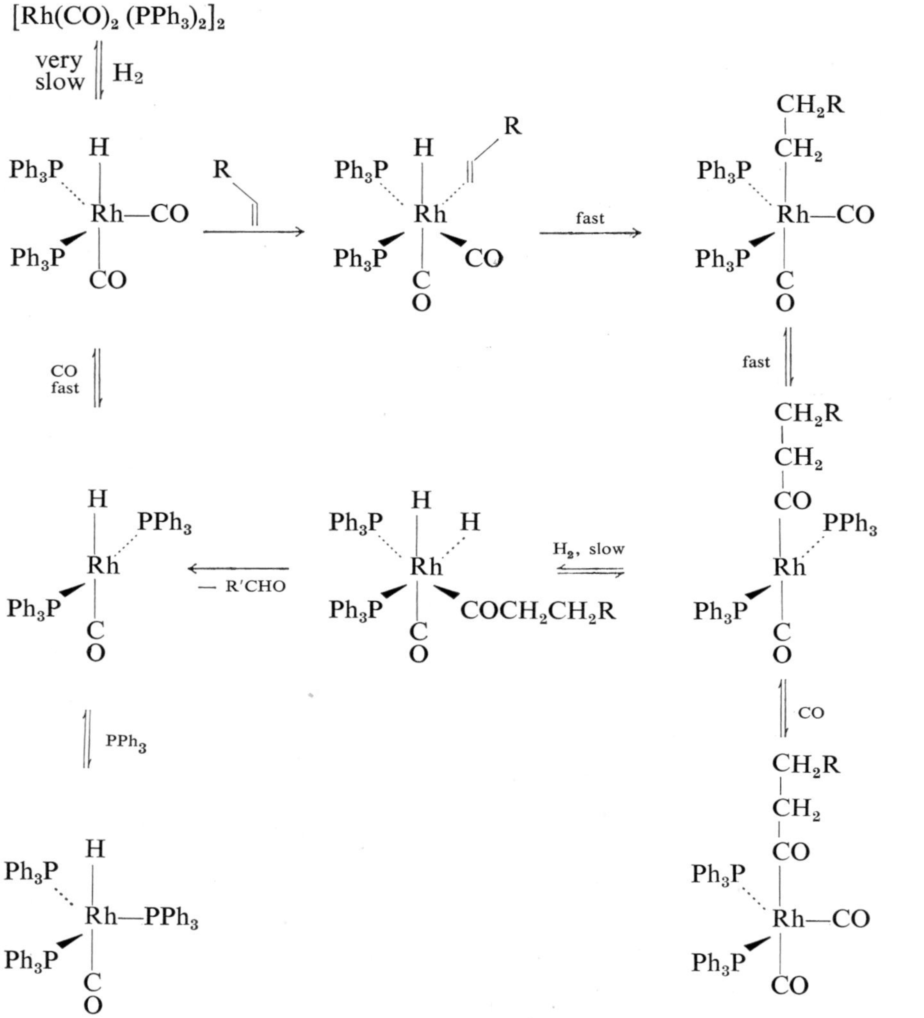

$[Rh(CO)_2(PPh_3)_2]_2$

very slow ‖ H_2

Figure 3. Proposed mechanism for the hydroformylation of olefins by the associative pathway [25].

carbonyl groups. In presence of triphenylphosphine the "red dimers" are quantitatively transformed by hydrogen at 25 °C and 1 atm to $RhH(CO)(PPh_3)_3$.

In a slow, but important reaction, solutions of both $RhH(CO)(PPh_3)_3$ and $[Rh(CO)(PPh_3)_2S]_2$ are transformed to $[Rh(CO)(PPh_3)_2]_2$ [113]. This dimer does not react with hydrogen at 25 °C and 1 atm, but only with carbon monoxide, producing $[Rh(CO)_2(PPh_3)_2]_2$. This fact explains, why the activity of $RhH(CO)(PPh_3)_3$ as an olefin hydrogenation [25, 73, 74] and isomerization

[25, 114] catalyst decreases with time: the dimer [Rh(CO) (PPh$_3$)]$_2$ is inactive as catalyst and inactive against hydrogen. In presence of carbon monoxide, on the other hand, no such deactivation of the catalyst can take place, since [Rh(CO) (PPh$_3$)]$_2$ is transformed through the "yellow dimer" again to RhH (CO) (PPh$_3$)$_3$. Accordingly, there is no decline in the activity of RhH(CO) (PPh$_3$)$_3$ as a hydroformylation catalyst.

Turning now to the mechanism of hydroformylation, two suggestions have been made by Wilkinson and co-workers [25], one of which has been termed the associative (Fig. 3.) and the other the dissociative (Fig. 4.) pathway. The main difference between the two alternatives lies in the mode of attack of the alkene on RhH(CO)$_2$ (PPh$_3$)$_2$, which is regarded as the actual catalyst: in the associative pathway this is a direct attack producing a 6 co-ordinated interme-diate, whereas in the dissociative pathway the first step is the formation of RhH (CO)$_2$ (PPh$_3$) by the loss of triphenyl phosphine and this 4 co-ordinate species is attacked by the olefin. Data are insufficient at present to decide between these two alternatives. The inhibiting effect of excess triphenylphosphine on hydrofor-mylation [25] can be interpreted in terms of both mechanisms.

The role of dicarbonyl complexes as key intermediates is supported by the observation that, if RhH(CO) (PPh$_3$)$_2$ and RhH(CO)$_2$ (PPh$_3$)$_2$ are present together in solution, only the latter complex reacts with ethylene at 25 ^{o}C as shown by the NMR spectrum [26].

Spectroscopic evidence has been found for the presence of acyl intermediates in the reaction of RhH(CO) (PPh$_3$)$_3$ with CO and terminal olefins: using ethylene at 1 atm total presure (1C$_2$H$_4$ + 1CO), complete conversion into a propionyl derivative, probably Rh(COEt) (CO)$_2$ (PPh$_3$)$_2$, was obtained [110, 111]. This species is stable in solution only in presence of carbon monoxide and ethylene and reacts with hydrogen (1 atm) to give propionaldehyde and RhH(CO) (PPh$_3$)$_3$:

$$\text{Rh(COEt) (CO)}_2 \text{ (PPh}_3)_2 \xrightarrow[\text{PPh}_3]{\text{H}_2} \text{RhH(CO) (PPh}_3)_3 + \text{EtCHO} + \text{CO}$$

The isolation of other intermediates has not been reported with simple olefins as substrates. Using tetrafluoroethylene more stable derivatives can be obtained [26], the structures of which are in agreement with the proposed mecha-nisms. These reactions with C$_2$F$_4$ are also shown in Fig. 2.

The effect of olefin structure on rate [11] is similar to that observed with unsubstituted rhodium carbonyls as catalysts [48] and suggests that mainly electronic and not steric factors influence reaction rate.

By adding an excess of triphenylphosphine to the hydroformylation reaction mixture the normal : iso aldehyde ratio is markedly increased [11]. This effect can be explained by both steric and electronic effects [25]. However, considering the effect of tributylphosphine in the case of methacrylic acid methyl ester [29], where it promotes the formation of the sterically more hindered α-formylisobutyric acid methyl ester, the electronic effect seems to predominate. This means that the incorporation of phosphine ligands instead of carbon monoxide into the

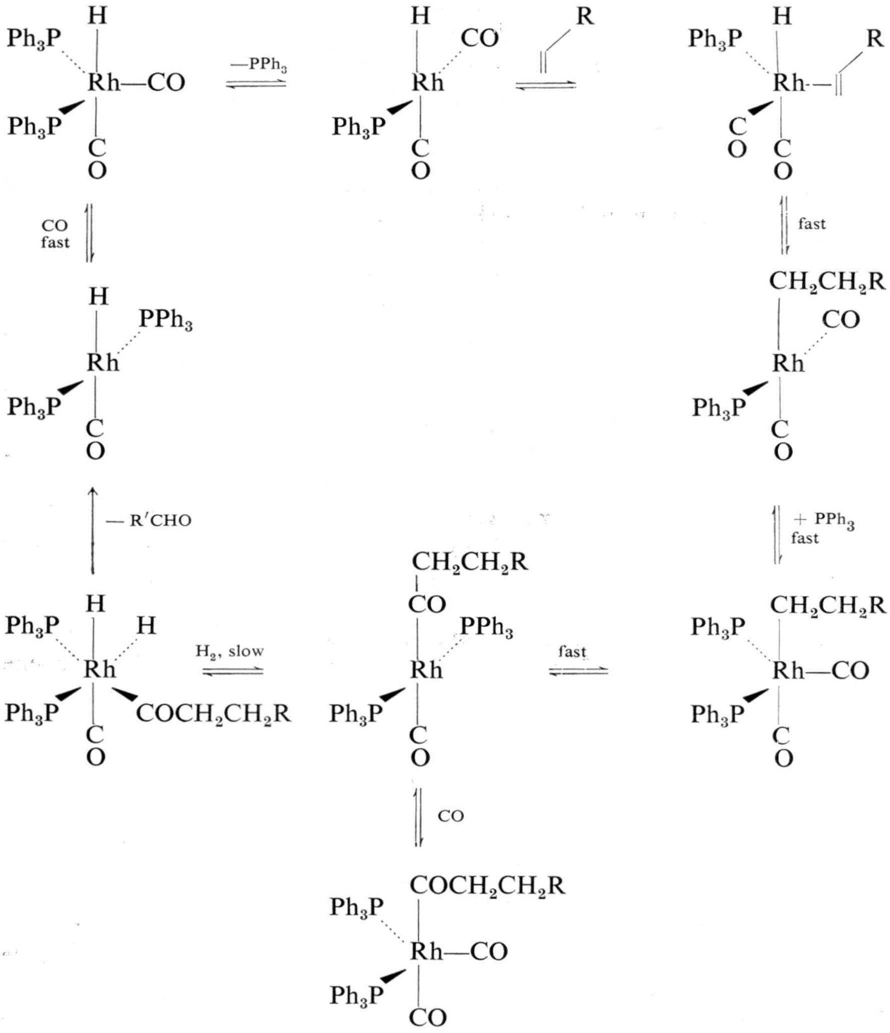

H
Ph₃P⋯Rh—CO   —PPh₃ ⇌   H CO   R   ⇌   Ph₃P H   R
Ph₃P   C   Rh   ‖   Rh⋯‖
          O   Ph₃P C   C C
                           O   O O

CO fast ⇅   fast ⇅

H PPh₃   CH₂CH₂R CO
Rh   Rh
Ph₃P C   Ph₃P C
      O   O

↑ — R'CHO   ⇅ + PPh₃ fast

H H   H₂, slow   CH₂CH₂R PPh₃   fast   CH₂CH₂R
Ph₃P⋯Rh   ⇌   CO   ⇌   Ph₃P⋯Rh—CO
Ph₃P C COCH₂CH₂R   Ph₃P Rh   Ph₃P C
      O   C   O
                            O

                            ⇅ CO

                            COCH₂CH₂R
                      Ph₃P⋯Rh—CO
                      Ph₃P CO

Figure 4. Proposed mechanism for the hydroformylation of olefins by the dissociative pathway [25].

catalytically active complex exerts its influence on the normal : iso ratio mainly by increasing the polarity of the rhodium-hydrogen bond in the sense $Rh^{\delta\oplus} — H^{\delta\ominus}$. This polarity favours an anti-Markovnikov type of addition of this bond to the double bond of the alkene. Olefinic hydrocarbons with terminal double bonds thus give rise mainly to straight chain aldehydes, whereas in the case of methacrylic acid methyl ester the branched product is favoured because of the opposite polarity of the olefinic double bond due to mesomeric forms such as:

The high PPh_3: Rh ratios needed to ensure specificity for the formation of the desired aldehyde isomer [11, 17, 29, 85] are obviously required in order to shift the equilibria between different CO- and PPh_3-containing complexes in favour of the species which contain more triphenylphosphine ligands. With respect to the associative and dissociative mechanisms proposed for hydroformylation this may mean that both are operative. The dissociative mechanism (intermediates containing fewer PPh_3 ligands) leads to a lower, and the associative one (intermediates containing more PPh_3 ligands) to a higher normal : iso aldehyde ratio [111].

3.7.3. HYDROFORMYLATION WITH OTHER COMPLEXES

Apparently no systematic studies corcerning the mechanism of hydroformylation with other rhodium complexes as catalysts have been performed.

In the case of the frequently used $Rh(CO)(PPh_3)_2Cl$ complex [17, 25, 50, 86], the reaction shows an induction period [25] which has been ascribed to transformation into a hydridic species. The formation of the hydride is facilitated if a base (e.g. triethylamine) is added to the reaction mixture and in this case no induction period is observed:

$$Rh(CO)(PPh_3)_2 Cl + H_2 \rightleftharpoons RhH(CO)(PPh_3)_2 + HCl$$

$$Rh(CO)(PPh_3)_2 Cl + H_2 + Et_3N \rightleftharpoons RhH(CO)(PPh_3)_2 + Et_3NHCl$$

Since the hydridic species formed in this process is the same of which is present in the experiments with $RhH(CO)(PPh_3)_3$ (Section 3.7.2), the mechanism of hydroformylation can be supposed to be the same as with the latter complex. Whether this is true also for the heterogeneous catalysts based on $Rh(CO)(PPh_3)_2Cl$ [78, 79, 86], cannot be decided. No data are available on the form that rhodium takes in those cases when $Rh(CO)(PPh_3)_2Cl$ is used and recirculated dissolved in a high boiling solvent [17, 50, 79].

The experience that the carboxylato complexes $[Rh(CO)_2 (OOCR)]_2$ are transformed under hydroformylation conditions to $Rh_4(CO)_{12}$ and $RCOOH$ [49] (Section 3.7.1.), also makes it appear likely that the catalysts of type $Rh(CO)(PR_3)_2 (OOCR')$ [8, 9, 10, 61, 62], $Rh(CO)L(AA')$ [61, 63] ($HAA' = \beta$-diketone), $Rh(CO) L (AN)$ [63] ($HAN = \beta$-diketone monoimine) and $Rh(CO)_2$ (amine) $(OOCR')$ [8, 10] yield hydridic rhodium carbonyl complexes and thus differ only in the way of forming the active catalysts. However, this supposition needs experimental confirmation.

The available data permit the conclusion that rhodium (I) complexes are in general much more active as hydroformylation catalysts than are rhodium (III) derivatives. Apparently this is a consequence of the fact that most of the intermediates of the catalytic cycles are hydrides, alkyls and acyls of Rh (I) (Fig. 3. and 4.).

3.8. Industrial Applications

No unambigous report on the application of a rhodium containing catalyst in an industrial hydroformylation process has appeared yet. Nevertheless, industry shows a pronounced interest in this field, as can be guessed from the numerous patents covering different aspects of the use of rhodium catalysts [1, 2, 3, 4, 6, 8, 9, 10, 16, 18, 23, 31, 33, 34, 37, 38, 43, 52, 62, 63, 70, 71, 77, 78, 79, 84, 88, 91, 99, 100, 107, 109]. Some of these will be mentioned briefly.

The main advantages of rhodium from the industrial standpoint are the possibility of producing aldehydes with special structures not accessible with cobalt catalysts and the formation of less hydrogenated by products if substituted olefins containing conjugated double bonds are reacted. Thus patents have been applied for the hydroformylation of N-acylamino olefins [1], carboxylic acid vinyl esters [2] styrene [99], acrylic esters [100], derivatives of unsaturated carboxylic acids [18] and methacrylic acid esters [31]. The hydroformylation of α-olefins to produce primary alchohols which are useful as starting materials for the manufacture of detergents [71] has been claimed.

Monsanto has developed two techniques for the hydroformylation of propylene with Rh(CO) (PPh$_3$)$_2$Cl as catalyst which avoid the need of catalyst regeneration [78, 79], one of the chief problems of industrial hydroformylation processes. One method uses a "gas sparged" reactor, which contains the catalyst dissolved in a high boiling solvent (dioctyl phthalate) [50, 79]. Reaction conditions (128 °C, 35 atm) are chosen to ensure that the butyraldehydes formed can leave the reactor with the excess of synthesis gas in the vapour phase. The catalyst, which is not volatile under these conditions, does not contaminate the product. The other method uses Rh(CO) (PPh$_3$)$_2$Cl on different types of supports as a heterogeneous catalyst [78, 79, 86].

The British Petroleum Co. has developed several new types of catalysts: the carboxylato complexes Rh(CO) (PR$_3$)$_2$ (OOCR') [8, 9, 10, 61, 62] and Rh(CO)$_2$ (amine) (OOCR') [8, 10], the diketonato complexes Rh(CO) (PR$_3$) (AA') [61, 63] (HAA' = β-diketone) and their derivatives Rh(CO) (PR$_3$) (AN) [63] (HAN = β-diketone monoimine). Some of these complexes are also active at atmospheric pressure, especially if alcohols [8] or N-disubstituted amides [62] are used as solvents.

The use of polar organic solvents instead of water prevents the formation and deposition of metallic rhodium [116].

Two methods have been patented for the recovery of rhodium from hydroformylation products [3, 33]. In general this seems to be a less cumbersome problem than in the case of cobalt. Under laboratory conditions usually a simple distillation of the product will suffice [32], since no volatile rhodium carbonyls are present and thus rhodium remains behind.

4. HYDROFORMYLATION WITH IRIDIUM COMPLEXES AS CATALYSTS

Iridium is a much less active catalyst for hydroformylation than either cobalt or rhodium mainly because of the greater stability of iridium complexes. Information on hydroformylation with iridium is therefore rather limited. Somewhat more work has been done on the chemistry of iridium hydrides and carbonyl complexes related to the supposed intermediates of the hydroformylation reaction.

The selectivity of iridium complexes in hydroformylation is generally poor and hydrogenation of the olefin is a significant concurrent reaction [5, 54, 55, 56, 57]. This is in agreement with the observation that many iridium complexes are good hydrogenation catalysts [53, 95, 103, 104].

Although no systematic studies have been made in this respect, the diminished activity seems to be characteristic of iridium catalysts irrespective of whether the metal is used in the form of "carbonyl" (probably $Ir_4(CO)_{12}$) [54, 55, 56, 57], oxide [32], metal on a support [88] or $Ir(CO)(PPh_3)_2Cl$ [25, 82]. This can also be seen from Tables 1,2 and 4. If iridium is applied as trichloride, the addition of tributylphosphine has a beneficial effect [92]. $IrH(CO)(PPh_3)_3$ is claimed to have the highest catalytic activity for hydroformylation among the Ir complexes; $IrH(CO)(PPh_3)_2Cl_2$ and $IrH_2(CO)(PPh_3)_2Cl$ give large amount of alkanes as byproducts [39].

The only iridium compound which has been investigated as a hydroformylation catalyst in somewhat more detail is $Ir(CO)(PPh_3)_2Cl$ [5]. Using this complex, 130-150 °C and 30 atm is needed to achieve 80-90% conversion of propylene in 1-4 hours. The yield of butyraldehydes was about 60% with a normal : iso ratio around 1.5 : 1, the remaining propylene was hydrogenated to propane. Hydrogenation was less if a lower temperature and hydrogen partial pressure were used. The addition of tributylphosphine had a beneficial effect on the normal : iso butyraldehyde ratio, but at the same time it strongly decreased the reaction rate.

The real catalysts are probably triphenylphosphine-substituted iridium carbonyl hydrides [25], but the addition of triethylamine to $Ir(CO)(PPh_3)_2Cl$ (which has a marked effect in the case of $Rh(CO)(PPh_3)_2Cl$) did not enhance the catalytic activity of the hydride but the catalytic activity of the latter which determines the reaction rate.

Complexes of the types IrH_3L_3 and IrH_3L_2 (L = PPh_3, PEt_2Ph) [53] have been claimed as catalysts for hydroformylation. It should be mentioned here

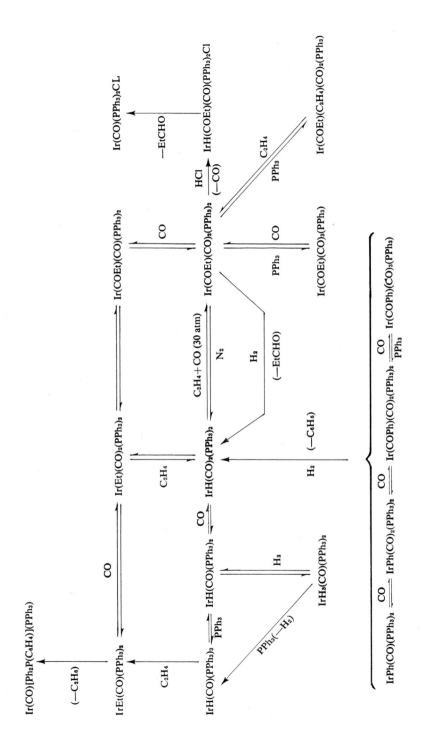

that, according to recent investigations, $IrH_3 (PEt_2Ph)_2$ is actually the pentahydride $IrH_5 (PEt_2Ph)_2$ and probably this is also the case with $IrH_3 (PPh_3)_2$ [68]. $IrH_3 (PPh_3)_3$ [15] and $IrH(CO)_2 (PPh_3)_2$ [111] are active catalysts for the isomerization of olefins.

Iridium complexes are often more stable than the corresponding rhodium compounds, which makes them very suitable as models for intermediates in catalytic processes. Some reactions of $IrH(CO)_2 (PPh_3)_2$, which can be regarded as the probable catalyst for hydroformylation, are shown in Fig. 5. These reactions have been helpful in elucidating the probable pathways of hydroformylation with rhodium complexes. Most of the transformations shown require no comment, only a few of them will be discussed briefly.

If $IrH(CO)_2 (PPh_3)_2$ is reacted with ethylene at 25 atm, the very labile $Ir (COEt) (CO) (PPh_3)_2$ can be isolated. If CO is applied too, the stable $Ir(COEt) (CO)_2 (PPh_3)_2$ is formed. This latter reacts with H_2 "much more slowly than its rhodium analogue" to give propionaldehyde and $IrH(CO_2) (PPh_3)_2$ [110, 111]:

$$Ir(COEt) (CO_2) (PPh_3)_2 \xrightarrow{H_2} IrH(CO_2) (PPh_3)_2 + EtCHO$$

This reaction closes the reaction cycle necessary for catalytic hydroformylation.

The reaction of $IrH(CO) (PPh_3)_3$ with ethylene yields ethane. This is explained by the loss of an *o*-hydrogen atom from one of the triphenylphosphine ligands [116]:

$$IrH(CO) (PPh_3)_3 \xrightarrow{C_2H_4} IrEt(CO) (PPh_3)_3 \longrightarrow$$

$$\longrightarrow Ph_2P\text{————}Ir(CO) (PPh_3)_2 + C_2H_6$$

The mixture of phenyl and benzoyl derivatives obtainable from $IrPh(CO)$ $(PPh_3)_2$ and CO reacts with hydrogen (in contrast to the propionyl complex) to yield benzene and not benzaldehyde [111]. This is also in contrast to the behaviour of the phenylrhodium and benzoylrhodium complexes formed from $RhPh(PPh_3)_3$ and CO, which yield benzaldehyde under similar conditions [110, 111] (Fig. 2.). This difference between analogous Ir and Rh complexes is in agreement with the general experience that Ir is rather a catalyst for hydrogenation than for hydroformylation.

5. HYDROFORMYLATION WITH RUTHENIUM AND OSMIUM COMPLEXES AS CATALYSTS

There are only two reports on the catalytic activity of osmium in hydroformylation [81, 82, 92]. Accordingly, $Os_3(CO)_{12}$ is only moderately active even at 180 °C and 160 atm ($1H_2 + 1CO$). Starting from propylene, equal amounts of normal and iso butyraldehydes were obtained (Tables 3. and 4.). Tributylphosphine is claimed to increase the catalytic activity of osmium [92].

The activity of $Ru_3(CO)_{12}$ as catalyst in hydroformylation of olefins appears to be definitely lower than that of either $CO(CO)_8$ or $Rh_4(CO)_{12}$. However, it is higher than that of osmium [81, 82] or iron carbonyls [25], which belong to the same group of the periodic system (Tables 3. and 4.). The rate of reaction is proportional to the olefin concentration. Carbon monoxide partial pressure also affects the rate of hydroformylation but has no influence on the normal : iso aldehyde ratio [81, 82]. A patent claims the use of metallic ruthenium, ruthenium salts or $Ru_3(CO)_{12}$ as catalysts for hydroformylation [94].

Phosphine complexes of ruthenium are also active as catalysts for hydroformylation and $Ru(CO)_3 (PPh_3)_2$ has been found to be the most effective. The following compounds were tested:

$Ru(CO)_3 (PPh_3)_2$	[24, 25]	$Ru(PPh_3)_2 Cl_3 \cdot CH_3OH$	[24, 25, 107]
$Ru(PPh_3)_4 Cl_2$	[24, 107]	$[Ru_2(PEt_2Ph)_6 Cl_3] Cl$	[107]
$Ru(PPh_3)_3 Cl_2$	[25, 107]		

A catalyst composed of $RuCl_3$, tributylphosphine and sodium acetate converted pentene-1 into a mixture of hexyl alcohols, mainly consisting of the normal isomer [91].

No information is available about the mechanism of these reactions.

6. HYDROFORMYLATION WITH PALLADIUM AND PLATINUM CATALYSTS

Ethylene (50 atm) and synthesis gas (100 atm) react at 100 °C in the presence of palladium chloride or metallic palladium to form propionaldehyde in low yield [101]. Ethane, formed by hydrogenation, is the main product. One patent describes the use of palladium chloride and diethylenetriamine for the hydroformylation of octene at 200 °C and 200 atm [6]. Aldehydes and alcohols were obtained in 38% yield.

The use of platinum complexes as catalysts for hydroformylation has only been described in patents [79, 92, 107]. Pt $(AsPh_3)_2Cl_2$ (at 70 °C and 85 atm) [107], $Pt(CO) (PBu_3)_2Cl$ (?) (at 195 °C and 35 atm) [79] and $PtCl_2 + PBu_3$ (at 195 °C and 30 atm) [92] were used; the activity was low.

7. HYDROGENATION OF ALDEHYDES
WITH CARBONYL DERIVATIVES
OF THE NOBLE METALS AS CATALYSTS

In sharp contrast to the homogeneous catalytic hydrogenation of olefins, for which a great number of highly active catalysts are known, the homogeneous hydrogenation of aldehydes is limited to only a few transition metal complexes [16, 75, 89, 90]. One of the chief difficulties is the decarbonylation of aldehydes which causes the formation of carbonyl complexes and thus leads to deactivation [60]. The hydrogenation of aldehydes to alcohols with rhodium and cobalt carbonyls as catalyst under conditions similar to that of olefin hydroformylation therefore deserves some attention. Since this review is confined to noble metals as catalysts, the reader should consult the book of Falbe [27] for information on aldehyde hydrogenation with cobalt carbonyls.

The hydrogenation of the aldehydes formed in hydroformylation with rhodium catalysts has often been seen to take place to a small extent. This consecutive reaction is obviously favoured by using longer reaction times [8, 63] or higher concentration of catalyst [9]. The addition of tributylphosphine (which in the case of cobalt carbonyls increases alcohol formation) has apparently no significant effect in this case [91, 93].

The most effective method for promoting aldehyde hydrogenation is by raising the temperature [75, 107], preferably to above 150 °C [28, 29, 30, 32, 44]. If the hydroformylation of olefins is performed at 170-230 °C alcohols are the main products, thus being obtained in a "one step" synthesis [44]. Generally it is more practicable, however, if the reaction is carried out in two steps by first performing hydroformylation under the usual conditions (80-160 °C) and raising the temperature (and usually also synthesis gas pressure) only after the olefin has been transformed to aldehydes [28, 29, 30, 32]. The main advantage of this method is that undesirable hydrogenation of the olefinic double bond, a side reaction which is also favoured by higher temperatures (Section 3.3.3.), can be avoided. Alcohol yields up to 90% are achieved.

The kinetics of aldehyde hydrogenation have been found [45] to follow (with n-butyraldehyde as a model substance and up to 70% aldehyde conversion) the rate law:

$$\frac{d\,(\text{alcohol})}{dt} = k\,[\text{aldehyde}]\,[\text{Rh}]^{1/6}\,(p_{H_2})^{0.5}\,(p_{CO})^{-0.3}$$

The reaction conditions were:

temperature 160 °C; p_{H_2} 20-180 atm; p_{CO} 80-160 atm

Below 80 atm carbon monoxide partial pressure the rate is proportional to p_{CO}, i.e. reaction rate shows a maximum at $p_{CO} = 80$ atm.

It was observed by infrared spectroscopy that rhodium which was added in the form of "RhCl$_3 \cdot$ 3H$_2$O" or Rh$_4$(CO)$_{12}$ is present in the reaction mixture mainly in the form of Rh$_6$(CO)$_{16}$. Based on this observation and on the kinetics equation above, the following reaction mechanism was proposed:

a) Formation of catalyst:

$$2 \, Rh_6(CO)_{16} + 4 \, CO \rightleftharpoons 3 \, Rh_4(CO)_{12}$$

$$Rh_4(CO)_{12} + 2 \, H_2 \rightleftharpoons 4 \, RhH(CO)_3$$

$$RhH(CO)_3 + CO \rightleftharpoons RhH(CO)_4$$

b) Catalytic reaction:

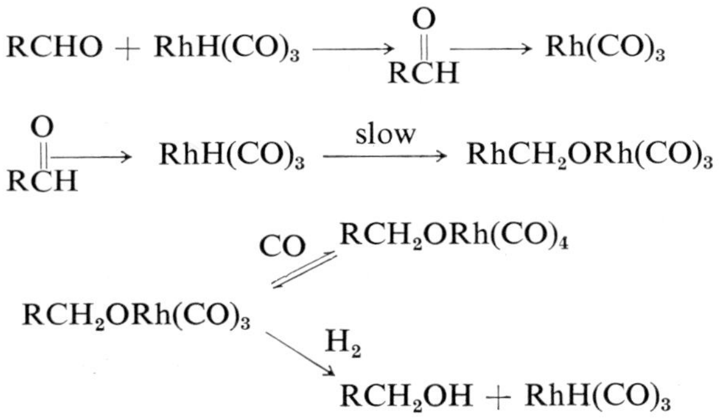

$$\text{RCHO} + \text{RhH(CO)}_3 \longrightarrow \underset{\text{RCH}}{\overset{\text{O}}{\underset{\|}{}}} \longrightarrow \text{Rh(CO)}_3$$

$$\underset{\text{RCH}}{\overset{\text{O}}{\underset{\|}{}}} \longrightarrow \text{RhH(CO)}_3 \xrightarrow{\text{slow}} \text{RhCH}_2\text{ORh(CO)}_3$$

$$\text{RCH}_2\text{ORh(CO)}_3 \overset{\text{CO}}{\underset{\text{H}_2}{\Bigg\langle}} \begin{array}{l} \text{RCH}_2\text{ORh(CO)}_4 \\ \\ \text{RCH}_2\text{OH} + \text{RhH(CO)}_3 \end{array}$$

The term $(p_{CO})^{-0.3}$ was explained by the supposition that the equilibrium between RhH(CO)$_3$ and RhH(CO)$_4$ is not shifted strongly to either side under the reaction conditions in distinct contrast to the similar equilibrium between CoH(CO)$_3$ and CoH(CO)$_4$ [102].

Falbe has successfully applied the combined hydroformylation and hydrogenation to produce special, usually bifunctional products from diolefins [28, 30] and substituted olefins [29, 30, 32]. Most of his results have already been quoted in Tables 11 and 12. Two aspects, not previously mentioned, will be discussed here.

If esters of unsaturated carboxylic acids are converted in this manner to the corresponding alcohols, the reaction goes further and the products which can be isolated are lactones [29, 31]. The formation of these lactones can be illustrated by the reaction sequences here reported.

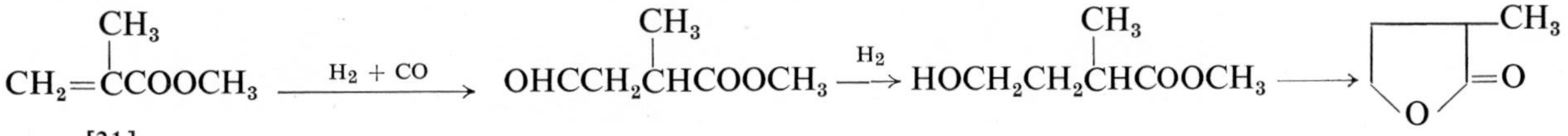

$$\underset{\text{[31]}}{\underset{\displaystyle\text{CH}_2=\overset{\displaystyle\text{CH}_3}{\text{C}}\text{COOCH}_3}{}} \xrightarrow{\text{H}_2+\text{CO}} \text{OHCCH}_2\overset{\text{CH}_3}{\text{CH}}\text{COOCH}_3 \xrightarrow{\text{H}_2} \text{HOCH}_2\text{CH}_2\overset{\text{CH}_3}{\text{CH}}\text{COOCH}_3 \longrightarrow$$

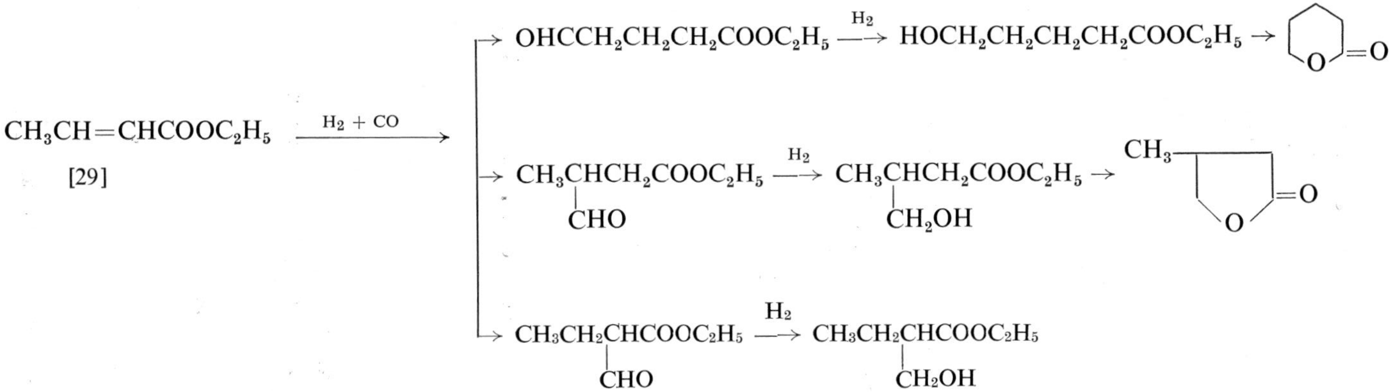

$$\underset{\text{[29]}}{\text{CH}_3\text{CH}=\text{CHCOOC}_2\text{H}_5} \xrightarrow{\text{H}_2+\text{CO}}$$

$$\text{OHCCH}_2\text{CH}_2\text{CH}_2\text{COOC}_2\text{H}_5 \xrightarrow{\text{H}_2} \text{HOCH}_2\text{CH}_2\text{CH}_2\text{CH}_2\text{COOC}_2\text{H}_5 \longrightarrow$$

$$\underset{\overset{\displaystyle|}{\text{CHO}}}{\text{CH}_3\overset{\displaystyle|}{\text{CH}}\text{CH}_2\text{COOC}_2\text{H}_5} \xrightarrow{\text{H}_2} \underset{\overset{\displaystyle|}{\text{CH}_2\text{OH}}}{\text{CH}_3\overset{\displaystyle|}{\text{CH}}\text{CH}_2\text{COOC}_2\text{H}_5} \longrightarrow$$

$$\underset{\overset{\displaystyle|}{\text{CHO}}}{\text{CH}_3\text{CH}_2\overset{\displaystyle|}{\text{CH}}\text{COOC}_2\text{H}_5} \xrightarrow{\text{H}_2} \underset{\overset{\displaystyle|}{\text{CH}_2\text{OH}}}{\text{CH}_3\text{CH}_2\overset{\displaystyle|}{\text{CH}}\text{COOC}_2\text{H}_5}$$

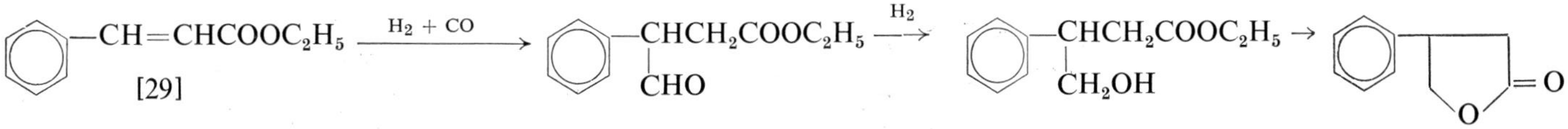

$$\underset{\text{[29]}}{\text{C}_6\text{H}_5\text{CH}=\text{CHCOOC}_2\text{H}_5} \xrightarrow{\text{H}_2+\text{CO}} \underset{\overset{\displaystyle|}{\text{CHO}}}{\text{C}_6\text{H}_5\overset{\displaystyle|}{\text{CH}}\text{CH}_2\text{COOC}_2\text{H}_5} \xrightarrow{\text{H}_2} \underset{\overset{\displaystyle|}{\text{CH}_2\text{OH}}}{\text{C}_6\text{H}_5\overset{\displaystyle|}{\text{CH}}\text{CH}_2\text{COOC}_2\text{H}_5} \longrightarrow$$

No lactones are formed of course if the hydroxymethyl group is in the α position with respect to the carboxylic group, as it is in the case of α-hydroxymethylbutyric acid ethyl ester which is derived from crotonic acid ethyl ester [29] (see scheme). The influence of temperature on product distribution has been discussed in Section 3.5. and can be seen in Table 16.

Minor amounts of formates are also formed from aldehydes by the hydroformylation of the carbon-oxygen double bond [49]:

$$RCHO + CO + H_2 \rightarrow RCH_2O-C\begin{array}{c}\nearrow O \\ \searrow H\end{array}$$

The formation of benzyl formate was observed when $Ru_3(CO)_{12}$ was prepared from Ru (III)-acetylacetonate in benzyl alcohol as solvent [81].

During the hydroformylation of propylene with $Ir(CO)(PPh_3)_2Cl$ as catalyst [5], only traces of butanols were formed.

The report of a patent [92], that pentene-1 is converted mainly to C_6 alcohols by catalysts composed of platinum, iridium or rhenium compounds and tributylphosphine (195 °C, 30 atm, $2H_2 + 1CO$), has not been confirmed yet.

8. RECENT RESULTS

The use of $RhH(CO)(PPh_3)_3$ as a catalyst in a continuous "oxo" process at 35 atm and 80-110 °C has been described [121]. The industrial applicability of the catalyst was proved by a 234 hours long run, in which 99,9% recovery of rhodium could be achieved. The kinetics and mechanism of hydroformylation with the same complex has been studied in detail [118] and the results fit the general scheme discussed in Section 3.6.3. Rate measurements make it probable, that the "associative pathway" (Fig. 3) is operative at high catalyst concentrations, while at low catalyst concentrations the "dissociative pathway" (Fig. 4) seems to be dominant. This catalytic system appears to be rather complex and even the numerous equilibria represented in Fig. 2, apparently do not completely characterize it. This is suggested for example by the observation, that the factors which influence the ratio of aldehyde isomers in the case of hexene-1 are not effective when styrene is the starting olefin.

The chemistry of iridium complexes related to hydroformylation has been extended by observing [120] the equilibria:

$$Ir_2(CO)_6(PPh_3)_2 \underset{-H_2}{\overset{+H_2}{\rightleftharpoons}} IrH(CO)_3(PPh_3) \underset{-H_2, +CO}{\overset{+H_2, -CO}{\rightleftharpoons}} IrH_3(CO)_2(PPh_3)$$

All three complexes are transformed into $Ir_4(CO)_9(PPh_3)_3$ by heating.

The catalytic properties of $IrH(CO)_3(PPh_3)$ have also been briefly reported [129].

The kinetics of hydroformylation has been studied with $Ru_3(CO)_{12}$ as catalyst [117]. The rate of reaction shows a maximum at relatively low carbon monoxide partial pressures ($\sim$ 20 atm at 150 °C), a feature which is very similar to that observed with rhodium or cobalt catalysts. At high temperature and low p_{CO}, hydrogenation of the olefin and the aldehyde became significant. The normal:iso aldehyde ratio is about 70% : 30% in the case of propylene, and is not influenced significantly by the reaction parameters. The amount of high boiling by products is less than 2%. No mechanism has been proposed for the reaction.

As shown by separate experiments $Ru_3(CO)_{12}$ is transformed at 150 °C by 100 atm H_2 into $Ru_4H_4(CO)_{12}$ [122], which is probably identical with $Ru_3H_4(CO)_{12}$, a complex obtained as byproduct in the polymethylene synthesis from H_2 and CO on solid ruthenium catalysts [124]. According to other investigations, however, two isomeric $Ru_4H_4(CO)_{12}$ complexes exist [119] and it can not be stated at present, which of these is formed from $Ru_3(CO)_{12}$ under conditions similar to hydroformylation. Furthermore, nothing can be said about the role of this tetrameric ruthenium hydride in the catalytic process, since it does not react with cyclohexene at 70 °C, but is "rather active in hydrogenation" [117].

It may be of interest to note in this connection, that in some products of of the polymethylene synthesis from carbon monoxide and hydrogen on ruthenium catalysts, ruthenium carbonyl derivatives (probably alkyl ruthenium carbonyls of high molecular weight) can be detected [123, 124].

Some rather interesting investigations have used high pressure infrared studies of species formed in situ at temperatures and pressures comparable to those used in conducting the reactions [125], or at very high pressure [126, 127].

The catalytic system formed in situ using $Rh(CO)(PPh_3)_2Cl$ [125] shows the following features: the complex $Rh(CO)(PPh_3)_2Cl$ is unaffected by CO/H_2 pressures (250 psi) at 80 °C however it is rapidly transformed under these conditions by addition of an olefin; the new rhodium species exhibit bands at 2021 and 2100 cm^{-1}. The addition of excess PPh_3 slightly prevents the rapid transformation of the rhodium complex.

Interestingly the rhodium complexes $HRh(CO)(PPh_3)_3$ and $HRh(CO)_2(PPh_3)_2$ are not stable at 80 °C and 500 psi of pressure CO/H_2; at reaction conditions the spectra observed when $Rh(CO)(PPh_3)_2Cl$ is used as catalyst precursor are completely different from the spectra observed when $HRh(CO)(PPh_3)_3$ is charged.

Whyman [126, 127] has done some high pressure studies on the existence of species such as $M_2(CO)_8$ (M = Rh, Ir) and $HM(CO)_4$ (M = Rh, Ir) and he has concluded that only $Rh_2(CO)_8$ and $IrH(CO)_4$ seem to exist in low concentration at very high pressure.

Some authours have also recently reported [128, 130, 131] on some rhodium catalysts where heterogeneous ligands are able to retain the transition metal within a polymeric matrix under conditions where retention may be difficult.

In the recent patent literature it was claimed that using high ratios ligand to rhodium (ligand=PPh₃ or P(OPh)₃; ratio 60-30) the rate of the hydroformylation of α-olefins is not too lowered and a very high ratio n to iso aldehydes is obtained [84]. The use of π-C₅H₅Rh(CO)PPh₃ as catalyst has also been claimed [132].

The use of Rh-carbonyl complexes of As, Sb, P containing ligands having As, Sb, P to N bonds (e.g. dipiperidino phenylphosphine) has been reported [133].

In a recent conference on the chemistry of hydroformylation and related reactions organised in Veszprem few interesting papers have been presented [134] such as hydroformylation of butadiene-1,3 and dicyclopentadiene using $trans$ RhCl(CO)(PPh₃)₂ and studies on the propylene hydroformylation with Rh₄(CO)₁₂ in mild conditions [135, 136].

Fell and Muller have reported also on the complex HRh(CO)(tribenzylamine)₃ and on the system rhodium-carbonyl-tertiary amine which achieves a direct hydroformylation of an olefin to alcohol at temperatures below 100 ºC and with high activity.

9. CONCLUSIONS

The development of noble metal catalysts has brought a tremendous change in the chemistry of hydroformylation. Looking at the amazing variety of catalysts available now and comparing this to the situation a few years ago, one must conclude that the real expansion of this field of chemistry is just beginning.

The large number of catalysts enables us to get a deeper insight into the mechanism of hydroformylation. To achieve this goal, more detailed kinetic measurements are needed; the catalysts active at atmospheric pressure may facilitate such investigations appreciably. In the case of rhodium, several parallel mechanisms seem to take place and a knowledge of these may render the control of product composition possible. The study of ruthenium (or osmium) catalysts will possibly lead to the recognition of a quite different type of hydroformylation mechanism, the result of which could perhaps be useful in developing catalysts based on iron.

Industry sooner or later will make use of a rhodium catalyst, first of all for the production of special products on a relatively small scale. Rhodium, due to its high activity and selectivity may open the field of organic intermediates (for the synthesis of pharmaceutics, dye stuffs, insecticides, etc.) to the hydroformylation process, which up till now is practically known only as a route to petrochemicals. The only serious problem seems to be posed by complete catalyst recovery, but once this is solved, a breakthrough can be achieved.

10. REFERENCES

[1] Ajinimoto Co., French Pat. 1.341.874 (1963); "Chem. Abstr.", *60*, 4041 (1964).

[2] Ajinimoto Co., French Pat. 1.361.797 (1964); "Chem. Abstr.", *61*, 11894 (1964).

[3] Ajinimoto Co., French Pat. 1.518.305 (1968); "Chem. Abstr.", *71*, 12665 (1969).

[4] J. H. BARTLETT and V. L. HUGHES (Esso Res. Eng. Co.), U.S. Pat. 2.894.038 (1959); "Chem. Abstr.", *54*, 2216 (1960).

[5] L. BENZONI, A. ANDREETTA, C. ZANZOTTERA and M. GAMIA, "Chimica e Industria" (Milan), *48*, 1076 (1966).

[6] H. W. BRADER, S. B. STANLEY and R. M. GIPSON (Jefferson Chem Co.), French Pat. 1.530.136 (1968); "Chem. Abstr.", *71*, 60714 (1969).

[7] D. S. BRESLOW and R. F. HECK, "Chem. Ind." (London), 467 (1960).

[8] Brit. Petr. Co., French Pat 1.549.414 (1968); "Chem. Abstr.", *72*, 2995 (1970). Neth. Pat. Appl. 67-14072 (1967); "Europ. Chem. News", *13* (333), 54 (1968).

[9] Brit. Petr. Co., French Pat. 1.558.222 (1969); "Chem. Abstr.", *72*, 31226 (1970).

[10] Brit. Petr. Co., French Pat. 1.573.158 (1969); Chem. Abstr. *72*, 100035 (1970).

[11] C. K. BROWN and G. WILKINSON, "Tetrahedron Lett.", 1725 (1969).

[12] S. H. H. CHASTON and F. G. A. STONE, "Chem. Comm.", 964 (1967); "J. Chem. Soc. A", 500 (1969).

[13] P. CHINI and S. MARTINENGO, "Inorg. Chim. Acta", *3*, 21 (1969).

[14] P. CHINI and S. MARTINENGO, "Chem. Comm.", 251 (1968); "Inorg. Chim. Acta", *3*, 315 (1969).

[15] R. S. COFFEY, "Tetrahedron Lett.", 3809 (1965).

[16] R. S. COFFEY, "Chem. Comm.", 923 (1967).

[17] J. H. CRADDOCK, A. HERSHMAN, F. E. PAULIK and J. F. ROTH, "Ind. Eng. Chem., Prod. Res. Develop.", *8*, 291 (1969).

[18] Deutsche Gold u. Silbersch. Anst., Neth. Pat. Appl. 65-16193 (1966); "Chem. Abstr.", *66*, 2215 (1967).

[19] M. DOKIYA and K. BANDO, "Kogyo Kagaku Zasshi", *71*, 1866 (1968).

[20] YA. T. EIDUS and A. L. LAPIDUS, "Neftehimiya", *7*, 51 (1967).

[21] YA. T. EIDUS, B. K. NEFEDOV, M. A. BESPROZVANNII and YU. V. PAVLOV, "Izv. Akad. Nauk SSSR, Ser. Him.", *7*, 1160 (1965).

[22] YA. T. EIDUS, B. K. NEFEDOV, M. A. BESPROZVANNII and YU. V. PAVLOV, "Neftehimiya", *6*, 282 (1966).

[23] J. L. EISEMANN (Diamond Alkali Co.), U.S. Pat. 3.290.379 (1966); "Chem. Abstr.", *66*, 46100 (1967).

[24] D. EVANS, J. A. OSBORN, F. H. JARDINE and G. WILKINSON, "Nature", *208*, 1203 (1965).

[25] D. EVANS, J. A. OSBORN and G. WILKINSON, "J. Chem. Soc. A", 3133 (1968).

[26] D. EVANS, G. YAGUPSKY and G. WILKINSON, "J. Chem. Soc. A", 2660 (1968).

[27] J. F. FALBE, *Synthesen mit Kohlenmonoxid*, (1967), Springer Verlag, Berlin.

[28] J. F. FALBE and N. HUPPES, "Brennstoff Chem.", *47*, 314 (1966).

[29] J. F. FALBE and N. HUPPES, "Brennstoff Chem.", *48*, 46 (1967).

[30] J. F. FALBE and N. HUPPES, "Brennstoff Chem.", *48*, 183 (1967).

[31] J. F. FALBE and N. HUPPES (Shell Oil Co.), U.S. Pat. 3.318.913 (1967); "Chem. Abstr.", *67*, 53715 (1967).

[32] J. F. FALBE, N. HUPPES and F. KORTE, "Brennstoff Chem.", *47*, 207 (1966).

[33] J. F. FALBE, H. TUMMES and J. MEIS (Ruhrchemie A.-G.), Ger. Pat. 1.295.537 (1969); "Chem. Abstr.", *71*, 51767 (1969).

[34] J. F. FALBE, H. TUMMES and J. WEBER (Ruhrchemie A.-G.), S. African Pat. 68-06828 (1969); "Chem. Abstr.", *72*, 43856 (1970).

[35] B. FELL and W. RUPILIUS, "Tetrahedron Lett.", 2721 (1969).

[36] B. FELL, W. RUPILIUS and F. ASINGER, "Tetrahedron Lett.", 3261 (1968).

[37] G. FOSTER, P. JOHNSON, M. J. LAWRENSON (Brit. Petr. Co.), Brit. Pat. 1.173.568 (1969); "Chem. Abstr.", *72*, 66388 (1970).

[38] G. FOSTER and M. J. LAWRENSON (Brit. Petr. Co.), Ger. Offen. 1.901.145 (1969); "Chem. Abstr.", 71, 123572 (1969).

[39] G. FOSTER and M. J. LAWRENSON (Brit. Petr. Co.), Ger. Offen. 1. 911.631 (1969); "Chem. Abstr.", *72*, 21322 (1970).

[40] V. YU. GANKIN, L. S. GENENDER and D. M. RUDKOVSKII, "Zh. Prikl. Him.", *40*, 2029 (1967).

[41] V. YU. GANKIN, L. S. GENENDER and D. M. RUDKOVSKII, "Zh. Prikl. Him.", *41*, 1577 (1968).

[42] V. YU. GANKIN, L. S. GENENDER and D. M. RUDKOVSKII, "Zh. Prikl. Him.", *41*, 2275 (1968).

[43] M. R. GIPSON (Jefferson Chem. Co.), Ger. Offen. 1.802.895 (1969); "Chem. Abstr.", *71*, 70067 (1969); Ger. Offen. 1.817.700 (1969); "Chem. Abstr.", *72*, 31228 (1970); Neth. Pat. Appl. 68-14533 (1969); "Europ. Chem. News", *16*. (392), 28 (1969).

[44] B. HEIL and L. MARKÒ, "Chem. Ber.", *99*, 1086 (1966).

[45] B. HEIL and L. MARKÒ, "Acta Chim. Acad. Sci. Hung.", *55*, 107 (1968).

[46] B. HEIL and L. MARKÒ, "Chem. Ber.", *101*, 2209 (1968).

[47] B. HEIL and L. MARKÒ, "Magyar Kém. Lapja", *23*, 669 (1968).

[48] B. HEIL and L. MARKÒ, "Chem. Ber.", *102*, 2238 (1969).

[49] B. HEIL and L. MARKÒ, "Chem. Ber." in press.

[50] A. HERSHMAN, K. K. ROBINSON, J. H. CRADDOCK and J. F. ROTH, "Ind. Eng, Chem., Prod. Res. Develop.", *8*, 372 (1969).

[51] W. HIEBER and H. LAGALLY, "Z. anorg. allg. Chem.", *251*, 96 (1943).

[52] V. L. HUGHES (Esso Res. Eng. Co.), U.S. Pat. 2.880.241 (1959); "Chem. Abstr.", *53*, 14398 (1959); Brit. Pat. 801.734 (1958); "Chem. Abstr.", *53*, 7014 (1959).

[53] Imperial Chem. Ind., Neth. Pat. Appl. 66-08122 (1966); "Chem. Abstr.", *67* 26248 (1967).

[54] N. S. IMYANITOV and D. M. RUDKOVSKII, "Neftehimiya", *3*, 198 (1963).

[55] N. S. IMYANITOV and D. M. RUDKOVSKII, *Oksosintez* (Gostoptehizdat, Leningrad), 30 (1963); "Chem. Abstr.", *60*, 9072 (1964).

[56] N. S. IMYANITOV and D. M. RUDKOVSKII, "Zhur. Prikl. Him.", *40*, 2020 (1967).

[57] N. S. IMYANITOV and D. M. RUDKOVSKII, "J. Prakt. Chem.", *311*, 712 (1969).

[58] B. R. JAMES and G. L. REMPEL, "J. Canad. Chem.", *44*, 233 (1966).

[59] F. H. Jardine, J. A. Osborn, G. Wilkinson and J. F. Young, "Chem. Ind." (London), 560 (1965).

[60] F. H. Jardine and G. Wilkinson, "J. Chem. Soc. C", 270 (1967).

[61] M. J. Lawrenson (Brit. Petr. Co.), Ger. Offen. 1.905.761 (1969); "Chem. Abstr.". 71, 124678 (1969). Neth. Pat. Appl. 69-01672 (1969); "Europ. Chem. News", 16, (407), 46 (1969).

[62] M. J. Lawrenson and G. Foster (Brit. Petr. Co.), Ger. Offen. 1.806.293 (1969); "Chem. Abstr.", 71, 70109 (1969).

[63] M. J. Lawrenson and G. Foster (Brit. Petr. Co.), Ger. Offen. 1.812.504 (1969); "Chem. Abstr.", 71, 101313 (1969); Neth. Pat. Appl. 68-17411 (1969); "Europ. Chem. News", 16, (399), 40 (1969).

[64] D. N. Lawson and G. Wilkinson, "J. Chem. Soc.", 1900 (1965).

[65] V. Macho, "Chem. Zvesti", 16, 73 (1962).

[66] I. Maeda and R. Yoshida, "Bull. Chem. Soc. Japan.", 41, 2969 (1968).

[67] L. Malatesta, G. Caglio and M. Angoletta, "J. Chem. Soc.", 6974 (1965).

[68] B. E. Mann, C. Masters and B. L. Shaw, "Chem. Comm.", 703 (1970).

[69] L. Markò, G. Bor, G. Almásy and P. Szabò, "Brennstoff Chem.", 44, 184 (1963).

[70] A. F. Millidge (Distillers Co.), French Pat. 1.411.602 (1965); "Chem. Abstr.", 64, 598 (1966).

[71] Mitsubishi Chem. Ind., Neth. Pat. Appl. 65-12800 (1966); "Chem. Abstr.", 65, 9186 (1966).

[72] M. Morikawa, "Bull. Chem. Soc. Japan", 37, 379 (1964).

[73] C. O'Connor and G. Wilkinson, "J. Chem. Soc. A", 2665 (1968).

[74] C. O'Connor, G. Yagupsky, D. Evans and G. Wilkinson, "Chem. Comm.", 420 (1968).

[75] J. A. Osborn, G. Wilkinson and J. F. Young, "Chem. Comm.", 17 (1965),

[76] J. A. Osborn, F. H. Jardine, J. F. Young and G. Wilkinson, " J. Chem. Soc. A", 1711 (1966).

[77] Y. H. Osumi, M. Yamaguchi, T. Onoda and M. Onishi (Mitsubishi Chem. Ind.), Jap. Pat. 22735 (1965); "Chem. Abstr.", 64, 4943 (1966).

[78] F. E. Paulik, K. K. Robinson and J. F. Roth (Monsanto Co.), U.S. Pat. 3.487.112 (1969); "Chem. Abstr.", 72, 68984 (1970).

[79] F. E. Paulik, J. F. Roth and K. K. Robinson (Monsanto Co.), French Pat. 1.560.961 (1969); "Chem. Abstr.", 72, 54762 (1970).

[80] F. Piacenti, P. Pino, R. Lazzaroni and M. Bianchi, "J. Chem. Soc. C", 488 (1966).

[81] P. Pino, G. Braca, F. Piacenti, G. Sbrana, M. Bianchi and E. Benedetti, Proceedings of New Aspects of the Chemistry of Metal Carbonyls and Derivatives (Venice, 2-4 sept. 1968), paper E 2.

[82] P. Pino, F. Piacenti, M. Bianchi and L. Lazzaroni, "Chimica e Industria" (Milan), 50, 106 (1968).

[83] P. Pino, F. Piacenti and P. P. Neggiani, "Chem. Ind.", (London), 1400 (1961).

[84] R. L. Pruett and J. A. Smith (Union Carbide Corp.), S. African Pat. 68-04937 (1968); "Chem. Abstr.", 71, 90819 (1969).

[85] R. L. Pruett and J. A. Smith, "J. Org. Chem.", 34, 327 (1969).

[86] K. K. ROBINSON, F. E. PAULIK, A. HERSHMAN and J. F. ROTH, "J. Catalysis", *15*, 245 (1969).

[87] S. SATO, M. TAKESADA and H. WAKAMATSU, "Nippon Kagaku Zasshi", *90*, 579 (1969).

[88] G. SCHILLER (Chem. Verwertungsges. Oberhausen), Ger. Pat. 953.605 (1956); "Chem. Abstr.", 11226 (1959).

[89] R. R. SCHROCK and J. A. OSBORN. " Chem. Comm.", 567 (1970).

[90] J. R. SHAPLEY, R. R. SCHROCK and J. A. OSBORN, "J. Amer. Chem. Soc.", *91*, 2816 (1969).

[91] L. H. SLAUGH and R. D. MULLINEAUX (Shell Oil Co.), U.S. Pat. 3.239.566 (1966); "Chem. Abstr.", *64*, 15745 (1966).

[92] L. H. SLAUGH and R. D. MULLINEAUX (Shell Oil Co.), U.S. Pat. 3.239.571 (1966); "Chem. Abstr.", *65*, 618 (1966).

[93] L. H. SLAUGH and R. D. MULLINEAUX, "J. Organometal. Chem.", *13*, 469 (1968).

[94] P. SMITH and H. H. JAEGER, (Imperial Chem. Ind.), Brit. Pat. 966.482 (1964); "Chem. Abstr.", *61*, 10593 (1964); Ger. Pat. 1.159.926 (1964); "Chem. Abstr.", *60*, 14389 (1964).

[95] W. STROHMEIER and T. ONODA, "Zeit. Naturforsch.", *24B*, 461 (1969).

[96] W. STROHMEIER and W. REHDER-STIRNWEISS, "Zeit. Naturforsch.", *24B*, 1219 (1969).

[97] W. STROHMEIER and W. REDHER-STIRNWEISS, "J. Organometal. Chem.", *19*, 417 (1969).

[98] Y. TAKEGAMI, Y. WATANABE and H. MASADA, "Bull. Chem. Soc. Japan", *40*, 1559 (1967).

[99] M. TAKESADA, H. WAKAMATSU, R. IWANAGA and J. KATO (Ajinimoto Co.), Jap. Pat. 1419 (1964); "Chem. Abstr.", *60*, 11950 (1964).

[100] M. TAKESADA, H. WAKAMATSU, R. IWANAGA and J. KATO (Ajinimoto Co.), Jap. Pat. 3020 (1964); "Chem. Abstr.", *60*, 15741 (1964).

[101] J. TSUJI, N. IWAMOTO and M. MORIKAWA, "Bull. Chem. Soc. Japan", *38*, 2213 (1965).

[102] F. UNGVÁRY and L. MARKÓ, "J. Organometal. Chem.", *20*, 205 (1969).

[103] L. VASKA, "Inorg. Nucl. Chem. Letters", *1*, 89 (1965).

[104] L. VASKA and R. E. RHODES, "J. Amer. Chem. Soc.", *87*, 4970 (1965).

[105] H. WAKAMATSU, "Nippon Kagaku Zasshi", *85*, 227 (1964).

[106] I. WENDER, H. W. STERNBERG and M. ORCHIN, "J. Amer. Chem. Soc.", *75*, 3041 (1953).

[107] G. WILKINSON, French Pat. 1.459.643 (1966); "Chem. Abstr.", *67*, 53652 (1967).

[108] G. WILKINSON, "Bull. Soc. Chim. France", 5055 (1968).

[109] G. WILKINSON, Neth. Pat. Appl. 68-01410 (1968); "Europ. Chem. News", *14*, (352), 36 (1968).

[110] G. YAGUPSKY, C. K. BROWN and G. WILKINSON, "Chem. Comm.", 1244 (1969).

[111] G. YAGUPSKY, C. K. BROWN and G. WILKINSON, "J. Chem. Soc. A", 1392 (1970).

[112] G. YAGUPSKY and G. WILKINSON, "J. Chem. Soc. A", 725 (1969).

[113] M. YAGUPSKY, C. K. BROWN, G. YAGUPSKY and G. WILKINSON, "J. Chem. Soc. A", 937 (1970).

[114] M. YAGUPSKY and G. WILKINSON, "J. Chem. Soc. A", 941 (1970).

[115] M. YAMAGUCHI, "Kogyo Kagaku Zasshi", 72, 671 (1969).

[116] M. YAMAGUCHI, T. ONODA, C. NAKAJIMA, M. KITATAMA and Y. YO, (Mitsubishi Chem. Ind.), Ger. Offen. 1.920.960 (1969); "Chem. Abstr.", 72, 31230 (1970).

[117] G. BRACA, G. SBRANA, F. PIACENTI and P. PINO, "Chimica e Industria" (Milan), 52, 1091 (1970).

[118] C. K. BROWN and G. WILKINSON, "J. Chem. Soc. A", 2753 (1970).

[119] B. F. JOHNSON, R. D. JONNSTON, J. LEWIS, B. H. ROBINSON and G. WILKINSON "J. Chem. Soc. A", 2856 (1968).

[120] L. MALATESTA, M. ANGOLETTA and G. CAGLIO, Proceedings XIIIth ICCC. Cracow-Zakopane, Sept. 14-22 (1970), Vol. 2., p. 338.

[121] K. L. OLIVIER and F. B. BOOTH, "Hydrocarbon Processing", (4), 112 (1970).

[122] F. PIACENTI, M. BIANCHI, E. BENEDETTI and P. FREDIANI, "Chimica e Industria" (Milan), 52, 81 (1970).

[123] H. PICHLER and B. FIRNHABER, "Brennstoff Chem.", 44, 33 (1963).

[124] H. PICHLER, H. MEIER zu KÖCKER, W. GABLER, R. GÄRTNER and D. KIOUSISS, "Brennstoff Chem.", 48, 266 (1967).

[125] D. E. MORRIS and H. B. TINKER, Abstracts Second North. Amer. Catalysis Soc. Meeting, Houston, U.S.A., Febr. 24-26 (1971).

[126] R. WHYMAN, "Chem. Comm.", 1381 (1969).

[127] R. WHYMAN, Proceedings, Reactivity and Bonding in Transition Organometallic Compounds, Venice, paper E2., (1970).

[128] W. O. HAAG and D. D. WHITEHURST, Abstracts Second North. Amer. Catalysis. Soc. Meeting, Houston, U.S.A., Febr. 24-26 (1971).

[129] L. MALATESTA, M. ANGOLETTA and F. CONTI, "J. Organomet. Chem.", 33, C43, (1971).

[130] W. O. HAAG and D. WHITEHURST (Mobil Oil Co.), Belgian Patent App., 721686.

[131] British Patent (B. P. Co), 70 06740 (1970).

[132] British Patent (B. P. Co), 70 08999 (1970).

[133] French Appl. (Inst. Français Pétrole), 2,041,776 (1970).

[134] Proceedings Chemistry of hydroformylation and related reactions, Veszprém, Hungary, May 31 - June 2nd (1972).

[135] B. L. BOOTH, M. J. ELSE, R. FIELDS and R. N. HASZELDINE, "J. Organomet. Chem", 27, 115 (1971).

[136] P. CHINI, S. MARTINENGO and G. GARLASCHELLI, "Chem. Comm", 709, (1972).

Chapter 2

Addition of Hydrogen Cyanide
to Mono-Olefins Catalyzed
by Transition Metal Complexes

E. S. BROWN

Union Carbide Corporation - South Charleston - U.S.A.

1. INTRODUCTION AND SCOPE

Alkyl nitriles are particularly valuable chemical intermediates serving as precursors to such diverse organic derivatives as carboxylic acids, esters, amides, amines and isocyanates. Nitriles may be synthesized by a variety of reactions including dehydration of carboxamides and aldoximes, dehydrogenation of amines, and nucleophilic displacement of halide by cyanide ion.

A synthetic method which offers wide applicability as well as simplicity is the direct addition of hydrogen cyanide to an olefin ("hydrocyanation"):

$$\ce{>C=C< + HCN -> >CH-C-CN}$$ (1)

The base catalyzed reaction [1, 2] is well-established and useful for the preparation of alkyl nitriles from olefins containing activating substituents, for example CN, COOR or COOH. Less well known is the hydrocyanation of non-activated as well as activated olefins occurring in the presence of certain transition metal complexes as catalysts [2]. Much of the work is descriptive in nature and the majority of the reports to date are found in the patent literature. As with other catalytic reactions, both homogeneous and heterogeneous vapor phase [3], catalyst systems have been developed, often in parallel. However, only the homogeneous catalytic reaction of mono-olefins is of concern here. The related oxycyanation [4] and alkyne hydrocyanation [5] reactions will not be discussed.

2. INTERACTION OF HYDROGEN CYANIDE WITH METAL COMPLEXES

Hydrogen cyanide is a reactive molecule which can interact with metal complexes (MeL_n) in two different ways:

$$MeL_n + HCN \longrightarrow L_nMe \leftarrow NCH$$ (2)

$$MeL_n + HCN \longrightarrow L_nMe \overset{H}{\underset{CN}{\diagdown}}$$ (3)

Reaction (2), where HCN behaves like a two electron donor, usually takes place with high oxidation state metal complexes which are effectively Lewis acids. Thus HCN adducts can be obtained with metal salts such as CuCl, $TiBr_4$, $SnCl_4$, $SbCl_5$, $FeCl_3$ or $BeCl_2$, $ZnCl_2$, BX_3 [38, 39].

Although the real nature of these adducts is not always clear, there is some evidence that coordination proceeds through the lone pair of the nitrogen atom of HCN.

Recently low oxidation state complexes of Mo and W of formula $Me(CO)_5HCN$ (Me = Mo, W) have been obtained [40]. In these complexes

HCN is nitrogen bonded to the metal; isomeric compounds of formula $Me(CO)_5CNH$, where the isomeric CNH acid is coordinated through the carbon atom, have also been reported [28].

Reaction (3), where HCN behaves like a protonic acid, takes place with low oxidation state transition metal complexes. Protonation of a metal complex by oxidative addition [41, 42] to form a transition metal hydride is a rather common reaction if homogeneous catalytic reactions of olefins by transition metals are considered. The metal-hydrogen bond is the active center of the catalytic process during insertion reactions with the formation of σ-alkyl intermediates [43]. In the case of reaction (3) the low oxidation state transition metal complexes behave as strong bases. Recently a great amount of work has been done on the Lewis basic properties of transition metal complexes [44].

The protonation reaction can give place to a side reaction in which hydrogen is evolved:

$$L_nMe \begin{matrix} H \\ \diagdown \\ CN \end{matrix} + HCN \longrightarrow L_nMe(CN)_2 + H_2 \qquad (4)$$

Reaction (4) which strongly depends on the solvent used, the concentration and the strength of the protonic acid, is related to the deactivation of the homogeneous catalysts.

3. HYDROCYANATION OF NON-ACTIVATED OLEFINS

Subsequent to a report of heterogeneous vapor phase hydrocyanation of non-activated olefins [6], Arthur, *et al* reported the activity of dicobalt octacarbonyl and derivatives as catalysts for the liquid phase hydrocyanation of a variety of non-activated olefins [7]. This appears to be the first instance of homogeneous catalysis. Later investigations have demonstrated that other cobalt complexes [8, 9] as well as complexes of ruthenium [10], rhodium [9], nickel [11-16], molybdenum, and tungsten [20] can function with varying degrees of effectiveness as hydrocyanation catalysts (see Table 1). It is interesting that with the exception of molybdenum and tungsten, the catalytically active metals reported are first and second row Group VIII metals.

Only limited effort has been made to determine the usefulness of many of the catalytically active metal complexes. For example, the complexes of ruthenium, rhodium, molybdenum, and tungsten have been employed only in the hydrocyanation of 3-pentene nitrile to adiponitrile. Arthur, *et al* reported the effectiveness of $Co_2(CO)_8$ with non activated olefins such as ethylene, terminal olefins and olefins having the bicyclo [2.2.1] hept-2-ene structure as well as such

Table 1

HOMOGENEOUS HYDROCYANATION OF NON-ACTIVATED MONOOLEFINS

Catalyst	Olefin	Product(s)	Ref.
$RuCl_2(PPh_3)_3$	$CH_3CH=CHCH_2CN$	$NC(CH_2)_3CN$, $CH_3CH(CN)(CH_2)_2CN$	[10]
$Co_2(CO)_8$	$CH_2=CH_2$	CH_3CH_2CN	[7]
	$CH_3CH=CH_2$	$CH_3CH(CN)CH_3$	[7]
	$CH_3CH_2CH=CH_2$	$CH_3CH_2CH(CN)CH_3$	[7]
	$CH_3CH=CHCH_3$	$CH_3CH(CN)CH_2CH_3$	[7]
	$CH_3(CH_2)_5CH=CH_2$	$CH_3(CH_2)_5CH(CN)CH_3$	[7]
	$CH_2=CH(CH_2)_3CN$	$CH_3CH(CN)(CH_2)_3CN$	[7]
	$CH_2=CH(CH_2)_3COOCH_3$	$CH_3CH(CN)(CH_2)_3COOCH_3$	[7]
	$CH_3CH=CHCH_2CN$	$CH_3CH(CN)CH_2CH_2CN$	[7]
	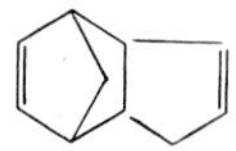	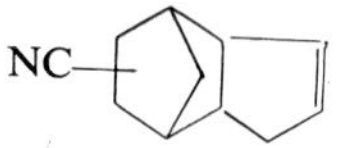	[7]
	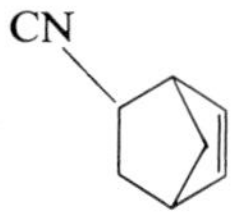	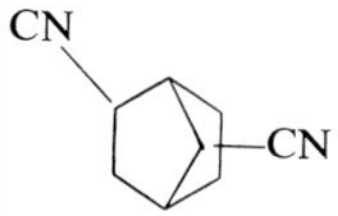	[7]
	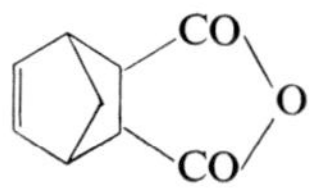	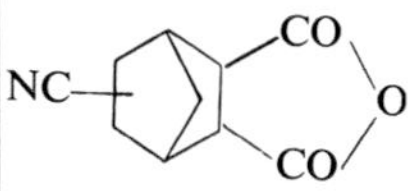	[7]

Catalyst	Olefin	Product(s)	Ref.
$Co_2(CO)_6(PPh_3)_2$	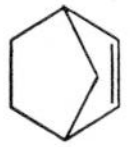 [bicyclic olefin]	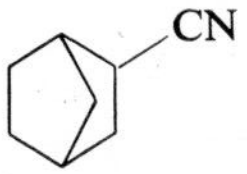 [bicyclic–CN]	[8]
$Co_2(CO)_6[P(OPh)_3]_2$	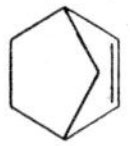 [bicyclic olefin]	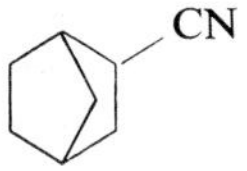 [bicyclic–CN]	[8]
$CoH[P(OPh)_3]_4$	$CH_3CH=CHCH_2CN$	$NC(CH_2)_4CN$, $CH_3CH(CN)(CH_2)_2CN$	[9]
	$CH_2=CHCH(CH_3)CN$	$NCCH_2CH_2(CH_3)CHCN$	[9]
Co^{2+} salts, reducing agent, $P(OR)_3$	$CH_3CH=CHCH_2CN$	$NCCH_2CH_2(CH_2)_2CN$	[9]
	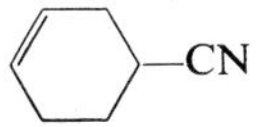 [cyclohexene–CN]	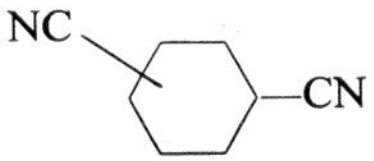 [cyclohexane NC–…–CN]	[9]
	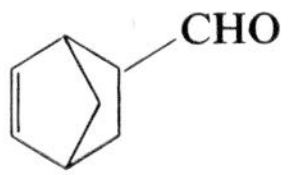 [bicyclic–CHO]	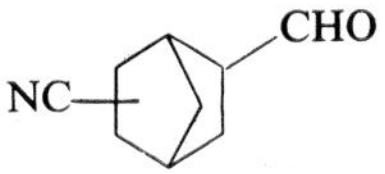 [bicyclic NC–…–CHO]	[9]
	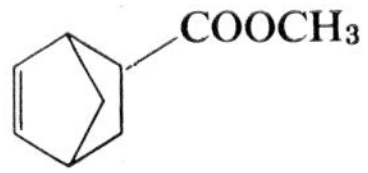 [bicyclic–COOCH₃]	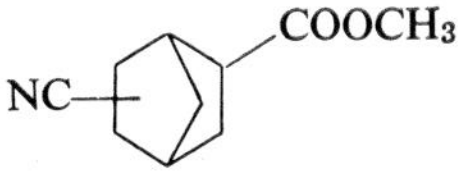 [bicyclic NC–…–COOCH₃]	[9]
$CoH[P(O-C_6H_4-CH_3)_3]_4$	$CH_2=CHCH_2CN$	$NC(CH_2)_3CN$	[9]
	$CH_3CH=CHCH_2CN$	$NC(CH_2)_4CN$	[9]

62

Table 1 (*Continued*)

HOMOGENEOUS HYDROCYANATION OF NON-ACTIVATED MONOOLEFINS

Catalyst	Olefin	Product(s)	Ref.
RhCl(PPh$_3$)$_3$	CH$_3$CH=CHCH$_2$CN	NC(CH$_2$)$_4$CN	[9]
Ni[P(OR)$_3$]$_4$, Zn^{2+} or Cd^{2+} salts	CH$_3$CH=CHCH$_2$CN	NC(CH$_2$)$_4$CN	[11]
NiL$_4$ [L = P(OR)$_3$, CO, PPh$_3$, AsPh$_3$, etc], BH$_4^-$	CH$_3$CH=CHCH$_2$CN	NC(CH$_2$)$_4$CN	[12, 13]
	CH$_2$=CHCH$_2$CN	NC(CH$_2$)$_3$CN, NCCH$_2$CH(CH$_3$)CN	[12, 13]
Ni[P(OR)$_3$]$_4$, P(OR)$_3$, metal salts	CH$_3$CH=CHCH$_2$CN	NC(CH$_2$)$_4$CN	[14]
	CH$_2$=CH—CH$_2$CN	NC(CH$_2$)$_3$CN, CH$_3$CH(CN)CH$_2$CN	[14]
	CH$_2$=CHCH$_2$CH$_2$CN	NC(CH$_2$)$_4$CN	[14]
	CH$_2$=CHCH(CH$_3$)CN	NCCH$_2$CH$_2$CH(CH$_3$)CN	[14]

	CH$_2$=C——CH$_2$
	        |
	    CH$_2$—CH—CN
		NC—CH$_2$—CH——CH$_2$
		           |
		        CH$_2$—CH—CN
			[14]

	C$_4$H$_9$CH=CH$_2$	C$_4$H$_9$CH$_2$CH$_2$CN(74%), C$_4$H$_9$CH(CN)CH$_3$(24%)	[14]
	 (cyclopentene)	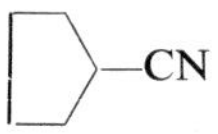 (cyclopentyl-CN)	[14]
	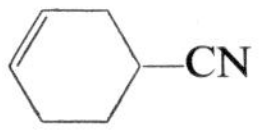 (cyclohexene-CN)	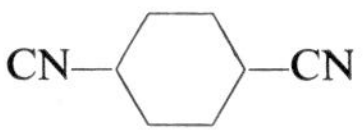 (cyclohexane-di-CN)	[13]
	C$_6$H$_{13}$CH=CH$_2$	C$_6$H$_{13}$CH$_2$CH$_2$CN(71%), C$_6$H$_{13}$CH(CN)CH$_3$(23%)	[14]

Catalyst	Olefin	Product(s)	Ref.
Ni[P(OR)₃]₄, P(OR)₃, metal salts	$CH_2{=}CH_2$	CH_3CH_2CN	[14]
	 (cyclopentene structure)	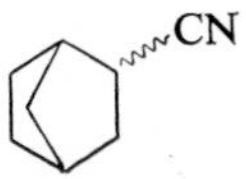 (cyclopentane–CN structure)	[14]
	$CH_3CH{=}CH{-}CH(CH_3)_2$	$(CH_3)_2CHCH_2CH(CN)(CH_3)$ $(CH_3)_2CHCH_2CH_2CH_2CN$	[14]
	$CH_3(CH_2)_2C(CH_3){=}CH_2$	$CH_3(CH_2)_2CH(CH_3)CH_2CN(53.1\%)$ $(CH_3)_2CH(CH_2)_3CN(36.4\%)$ $(CH_3)_2CHCH_2CH_2CH(CN)CH_3(6.7\%)$	[14]
	cis/trans $CH_3(CH_2)_2CH{=}CHCH_3$	$CH_3(CH_2)_5CN(75\%)$, $CH_3(CH_2)_3CH(CN)CH_3(25\%)$	[14]
Ni[P(OR)₃]₄, P(OR)₃, metal salts or boron compounds	$CH_3CH{=}CHCH_2CN$	$NC(CH_2)_4CN$	[15]
Ni²⁺ salt, Zn or Cd powder, P(OR)₃	$CH_3CH{=}CHCH_2CN$	$NC(CH_2)_4CN$	[16]
Pd[P(OR)₃]₄, metal salts or boron compounds	$CH_3CH{=}CHCH_2CN$	$NC(CH_2)_4CN$	[17]
	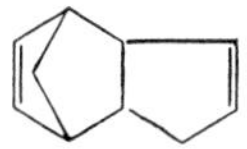 (dicyclopentadiene structure)	hydrocyanated dicyclopentadiene	[17]
Pd[P(OPh)₃]₄, P(OPh)₃	 (bicyclic olefin structure)	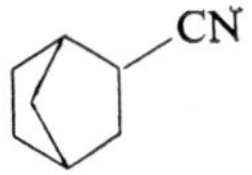 (bicyclic–CN structure)	[18, 19]

Table 1 (*continued*)

HOMOGENEOUS HYDROCYANATION OF NON-ACTIVATED MONOOLEFINS

Catalyst	Olefin	Product(s)	Ref.
Mo[P(OPh)₃]₃(CO)₃, TiCl₃	(norbornene derivative, NC)	NC–/–CN product	[19]
Mo[P(O-⟨C₆H₄⟩-CH₃)₃]₃(CO)₃	(norbornene derivative, NC)	NC–/–CN product	[19]
BPh₃, P(O-⟨C₆H₄⟩-CH₃)₃		–CN (86%) / –CN (16%)	[19]
			[19]
			[19]
	CH₃CH=CHCH₂CN	NC(CH₂)₄CN	[20]
	CH₃CH=CHCH₂CN	NC(CH₂)₄CN	[20]
W[P(OPh)₃]₃(CO)₃, BPh₃	CH₃CH=CHCH₂CN	NC(CH₂)₄CN	[20]

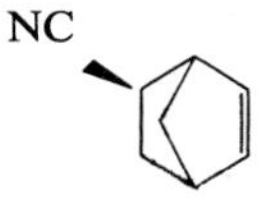

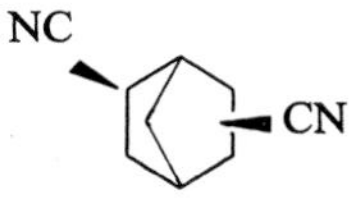

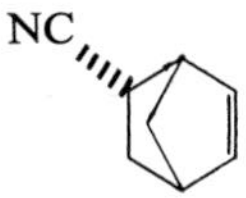

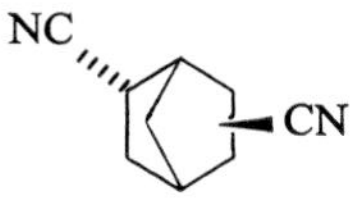

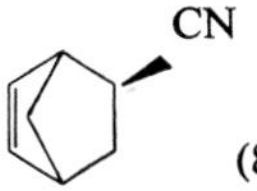

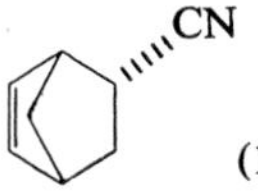

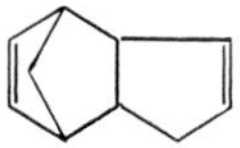

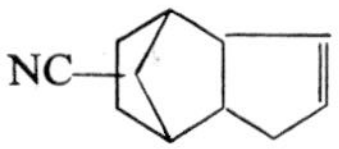

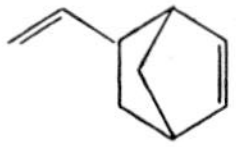

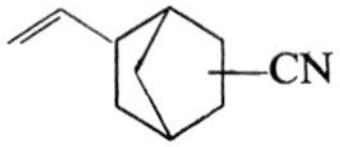

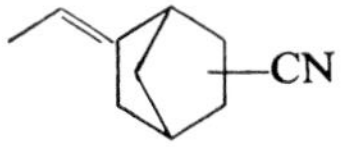

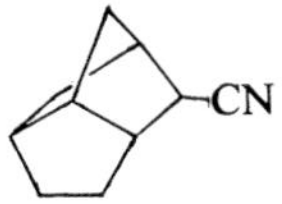

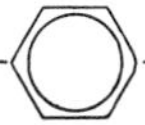

activated olefins as styrene. Hydrocyanation of these same types of olefins has been noted for many of the other catalysts (Table 1). Olefins having internal double bonds generally react only sluggishly; this fact is typical of many addition reactions catalyzed by transition metal complexes.

The behavior of strained olefins has been investigated in some depth. Bicyclo [2.2.1] hept-2-ene reacts with HCN in the presence of Pd[P(OPh)$_3$]$_4$, for example, to form *exo*-5-cyanobicyclo-[2.2.1] heptane [18]. No *endo*-5-cyanobicyclo [2.2.1] heptane was observed in the reaction product:

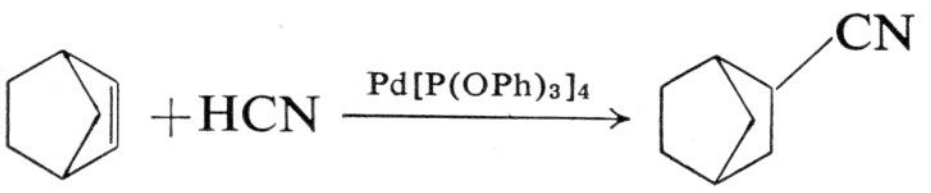

$$\tag{5}$$

Whether the reaction involves *cis* or *trans* addition of HCN has not yet been established.

Similarly, *exo*-5-cyanobicyclo[2.2.1]hept-2-ene forms dinitrile products in the presence of Co$_2$(CO)$_8$ [7] or zerovalent palladium catalysts [19]:

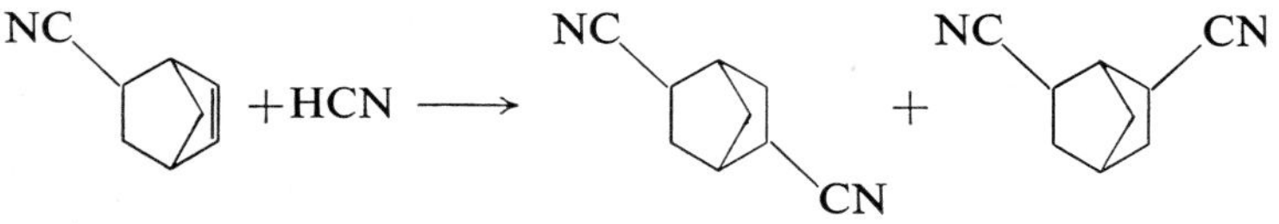

$$\tag{6}$$

Again, use of Pd[P(OPh)$_3$]$_4$ affords products having *exo*-steric configuration of the entering cyano substituent. Similarly, when *endo*-5-cyanobicyclo[2.2.1]-hept-2-ene is reacted with HCN only *exo*-(5 or 6)-cyano-*endo*-2-cyanobicyclo-[2.2.1]heptane is obtained.

An interesting reaction of this type is the palladium catalyzed addition of HCN to 5-vinylbicyclo[2.2.1]hept-2-ene [19]. Under the conditions employed not only hydrocyanation of the strained double bond is observed but isomerization of the vinyl substituent occurs to some extent:

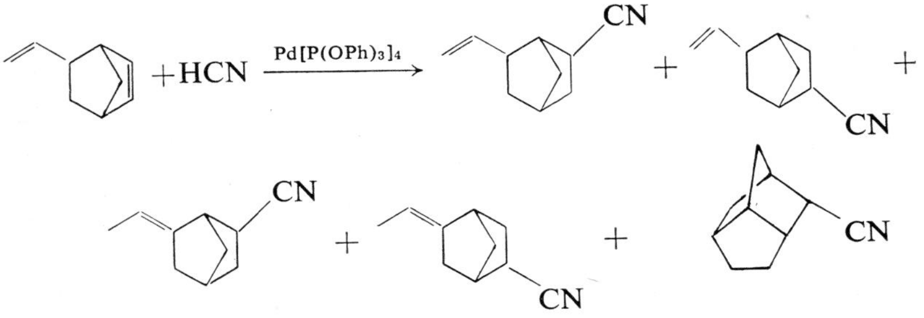

$$\tag{7}$$

Formation of *endo*-2-cyanotricyclo [4.2.1.0] nonane (cyanobrendane) [3, 7] may be rationalized by assuming that *endo*-5-vinylbicyclo[2.2.1]-hept-2-ene present in the starting olefin mixture forms a chelate complex providing a pathway by which carbon-carbon bond formation can accompany hydrocyanation:

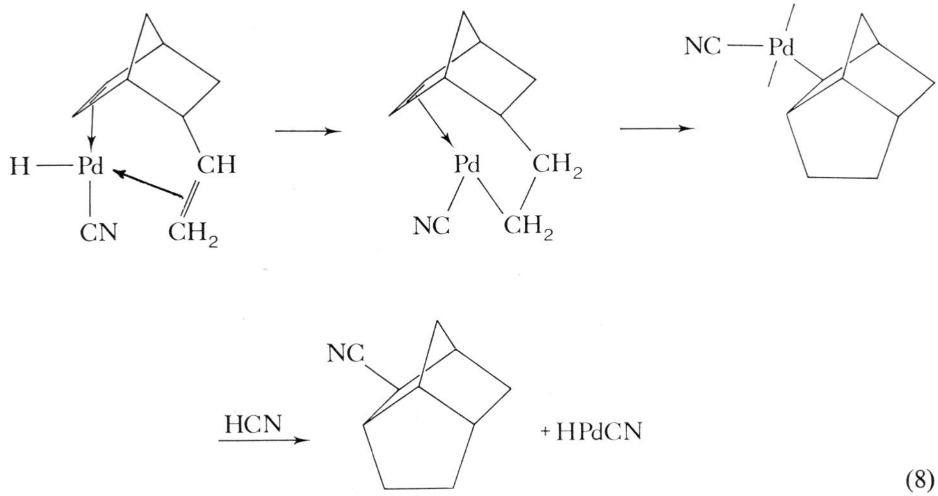

$$(8)$$

As might be predicted, only the strained double bond of dicyclopentadiene reacts with HCN in the presence of either $Co_2(CO)_8$ or $Pd[P(OPh)_3]_4$ catalysts [7, 19]:

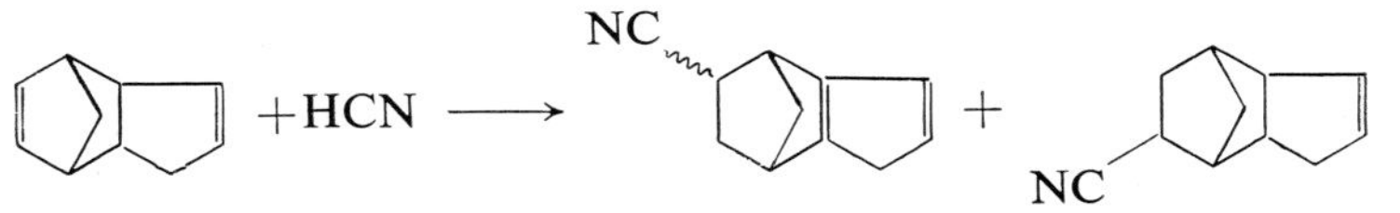

$$(9)$$

As noted earlier, olefin isomerization may occur under hydrocyanation conditions and may prove a source of difficulty, especially in view of the possible reactivity differences between terminal and internal double bonds. This is turned to an advantage, however, in the reaction of 3-pentenenitrile where isomerization to 4-pentenenitrile precedes hydrocyanation to afford adiponitrile as the major product [11, 12]. This reaction has received considerable attention because it provides the final step in a potential industrial synthesis of adiponitrile from butadiene. This reaction will not be however treated in detail here.

$$CH_2{=}CH{-}CH{=}CH_2 + HCN \longrightarrow CH_3CH{=}CH{-}CH_2CN \qquad (10)$$

$$CH_3CH{=}CH{-}CH_2CN \longrightarrow CH_2{=}CH{-}(CH_2)_2CN \qquad (11)$$

$$CH_2{=}CH(CH_2)_2CN + HCN \longrightarrow NC(CH_2)_4CN \qquad (12)$$

Ethylene reacts with HCN to form propionitrile in the presence of $Co_2(CO)_8$ [7] or $Pd[P(OPh)_3]_4$ [18]:

$$CH_2{=}CH_2 + HCN \longrightarrow CH_3CH_2CN \qquad (13)$$

Hydrocyanation of propylene in the presence of $Co_2(CO)_8$ is reported to yield to isobutyrronitrile exclusively [7]:

$$CH_3CH{=}CH_2 + HCN \xrightarrow{Co_2(CO)_8} CH_3CH(CN)CH_3 \qquad (14)$$

When this reaction is performed using $Pd[P(OPh)_3]_4$, however, a mixture of isomers is formed [8]:

$$CH_3CH{=}CH_2 + HCN \xrightarrow{Pd[P(OPh)_3]_4} \underset{75\%}{CH_3CH_2CH_2CN} + \underset{25\%}{CH_3CH(CN)CH_3} \qquad (15)$$

The structure of the reaction products is dependent upon thee factors: the catalyst employed, the solvent and the temperature of the reaction. For example, the predominance of Markownikoff addition products appears to be general for the $Co_2(CO)_8$ catalyzed reactions of terminal olefins [7], whereas anti-Markownikoff addition products appears to be favored in the reactions catalyzed by zerovalent nickel and palladium:

$$CH_2{=}CH(CH_2)_3CN + HCN \xrightarrow{Co_2(CO)_8} CH_3CH(CN)(CH_2)_3CN \qquad (16)$$

$$CH_2{=}CH(CH_2)_2CN + HCN \xrightarrow{Ni[P(OC_2H_5)_3]_4} \underset{85\%}{NC(CH_2)_4CN} +$$

$$\underset{15\%}{CH_3CH(CN)(CH_2)_2CN} \qquad (17)$$

As we will see later the presence of a Lewis acid cocatalyst increases the yields in adiponitrile starting either from 3- or 4-pentenenitrile or also from 2 methyl,3-butenenitrile [20, 51, 54, 55, 56).

Only two instances of HCN addition to cyclic olefins have been reported; formation of cyanocyclopentane from cyclopentene and preparation of 1,4-dicyanocyclohexane from 4-cyano-1-cyclohexene:

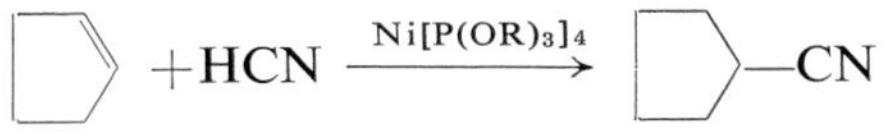

$$(18)$$

$$(19)$$

In summary, catalytic activity for hydrocyanation of non-activated mono olefins has been demonstrated for a number of transition-metal complexes, notably those zerovalent of Co, Ni and Pd.

With the exception of catalysis by $Co_2(CO)_8$ and Ni(O) complexes the number of olefin types hydrocyanated by the various metal complexes is curiously limited. Whether this reflects serious restrictions on the utility of the reaction or is simply the result of limited effort to date remains to be determined.

4. THE PROPOSED MECHANISMS OF HYDROCYANATION

The mechanism of hydrocyanation is not known, but there is suggestive evidence that the reaction proceeds by a sequence of the following type which has, as its starting point, the oxidative addition of HCN to the metal complex (ligands are omitted for sake of simplicity):

$$M + HCN \rightleftharpoons H—M—CN \qquad (20)$$

$$H—M—CN + \text{olefin} \longrightarrow H—M(\pi\text{-olefin})(CN) \qquad (21)$$

$$H—M(\pi\text{-olefin})(CN) \longrightarrow \sigma\text{-alkyl}—M—CN \qquad (22)$$

$$\sigma\text{-alkyl}—M—CN \longrightarrow \text{alkyl}—CN + M \qquad (23)$$

Oxidative additions of HCN to phosphine complexes of Rh(I) [21], Ir(I) [21, 22], and Pt(O) [23d] have been reported. The addition of DCN to Ir(I) [21, 22] is also known:

$$RhCl(PPh_3)_3 + 2HCN \longrightarrow \begin{array}{c} Ph_3P \quad H \quad CN \\ \diagdown \quad | \quad \diagup \\ Rh \\ \diagup \quad | \quad \diagdown \\ (HCN) \quad Cl \quad PPh_3 \end{array} \qquad (24)$$

$$IrCl(CO)(PPh_3)_2 + HCN \longrightarrow \begin{array}{c} PPh_3 \quad CN \quad H(D) \\ \diagdown \quad | \quad \diagup \\ Ir \\ \diagup \quad | \quad \diagdown \\ OC \quad Cl \quad PPh_3 \end{array} \qquad (25)$$

$$Pt(PPh_3)_3 \text{ or } _4 + HCN \longrightarrow trans\text{- } Pt(H)(CN)(PPh_3)_2 \qquad (26)$$

Since neither complexes of Pt(O) nor Ir(I) have been reported effective as hydrocyanation catalysts, it is likely that instability of the hydrocyanide complex or, more probably, that of the postulated alkylcyano complex formed subsequently is crucial to effective catalysis (Note that *trans*-Pt(C$_2$H$_5$)(CN)[P(C$_2$H$_5$)$_3$]$_2$ has been prepared, but there is no evidence of its forming propionitrile on thermolysis [24]).

Tris(triphenylphosphine)chloro rhodium(I) which forms an isolable HCN adduct hydrocyanates 3-pentenenitrile to produce adiponitrile and α-methylglutaronitrile (see Table 1). The reaction of Ni[P(OC$_2$H$_5$)$_3$]$_4$ in solution with HCN among other acids has been followed by observing the high field resonance due to hydrides (23a,b,c 25]. From the nmr splitting pattern, it appears that a pentacoordinate hydride NiH[P(OC$_2$H$_5$)$_3$]$_3$CN is formed which is in equilibrium with a tetracoordinate hydride NiH[P(OC$_2$H$_5$)$_3$]$_2$CN, The behaviour of Pd(O) complexes, with HCN alone does not appear to have been fully investigated. By analogy, in their reaction with aqueous acids to form M(II) salts, however, these complexes would not be expected to form stable adducts [23d].

Only recently the formation of Pd(II) hydrides by protonation of Pd(O) carbonyl phosphine complexes with HX has been reported [45].

The involvement of a metal hydride in hydrocyanation is given credence by the catalytic activity of CoH(PR$_3$)$_4$ complexes. Hydride involvement is also suggested by the frequent occurrence of olefin isomerization during hydrocyanation.

One difficulty with the proposed mechanism lies in the formation of alkyl nitrile by reductive elimination (equation (23)). There is no precedent for such a reductive elimination.

There is evidence suggesting that perhaps alternative pathway may be operating. Equation (23) may be rewritten as:

$$\sigma\text{-alkyl—M—CN} + \text{HCN} \longrightarrow \text{alkyl—CN} + \text{H—M—CN} \qquad (23a)$$

summarizing a sequence in which the coordinated cyanide ligand is first protonated to form coordinated hydrogen isocyanide [28-30]. This ligand is then "inserted" into alkyl metal bond to form an alkylimino-metal complex:

$$\sigma\text{-alkyl—M—CN} + \text{HCN} \longrightarrow \sigma\text{-alkyl—M(HCN)CN} \qquad (27)$$

$$\sigma\text{-alkyl—M(HNC)CN} \longrightarrow \text{alkyl—}\overset{\overset{\displaystyle NH}{\|}}{C}\text{—M—CN} \qquad (28)$$

$$\text{alkyl—}\overset{\overset{\displaystyle NH}{\|}}{C}\text{—MCN} \longrightarrow \text{alkyl—CN} + \text{H—MCN} \qquad (29)$$

which decomposes to form the alkyl nitrile and regenerates the metal hydrido-cyanide complex. This sequence is suggested by the mechanism of the formation of alkyl nitriles from alkylpentacyanocobaltate (III) first proposed by Kwiatek and Seyler [26] and subsequently substantiated by the kinetic studies of Tobe *et al* [27].

Moreover recent studies [46] have shown that carbon bonded isonitriles RNC can insert into the alkyl-nickel bond of π-C$_5$H$_5$Ni(PPh$_3$)R′ to give π-C$_5$H$_5$Ni(CNR)[C(R′) = NR].

Hydrocyanations catalyzed by $Co_2(CO)_8$ suffer the limitation that large quantities of the complex are required to achieve satisfactory olefin conversions. Dicobalt octocarbonyl reacts with HCN alone to form a blue compound. This compound contains CN and CO ligands and when treated with aqueous KCN, forms $K_3 [Co(CN)_6]$ liberating H_2, CO, and HCN [7]. The blue compound, although less effective than $Co_2(CO)_8$ itself, does catalyze olefin hydrocyanation but rapidly declines in activity.

Similar deactivating processes occur in hydrocyanations catalyzed by zerovalent nickel and palladium complexes. It is reported that in the course of hydrocyanation of 3-pentenenitrile catalyzed by tetrakis[triphenylphosphite]-nickel, insoluble nickel salts (corresponding to about 60% of the added nickel) precipitate primarily on $Ni(CN)_2$ (Nickel cyanide is catalytically inactive under the reaction conditions). If, however, the same reaction is conducted in the presence of excess triphenylphosphite (6 moles per mole of nickel complex), a slower rate of hydrocyanation results, but only about 20% of the $Ni[P(OPh)_3]_4$ has been converted to $Ni(CN)_2$. Although the reaction rate is slowed, presumably the catalyst life is prolonged in this manner.

A similar deactivation process has been observed during hydrocyanations with $Pd[P(OPh)_3]_4$. As in the case of nickel, loss of catalytic activity can be correlated with the formation of palladium(II) cyanide complexes [18]. The deactivation process may be suppressed by employing excess phosphorus containing ligand, but at the expense of a reduction in the rate of hydrocyanation.

At present the best explanation for these observations is that competitive with the hydrocyanation process, either the intermediate hydrido metal cyanide or the alkyl metal cyanide reacts further with HCN to form an unstable metal(IV) dicyanide complex which subsequently decomposes giving metal(II) cyanide, and either hydrogen [31] or alkane, respectively:

$$H\!-\!M\!-\!CN + HCN \longrightarrow [H_2M(CN)_2] \longrightarrow M(CN)_2 + H_2 \qquad (30)$$

$$\sigma\text{-alkyl}\!-\!M\!-\!CN + HCN \rightarrow [\sigma\text{-alkyl}\!-\!M(H)(CN)_2] \rightarrow M(CN)_2 + \text{alkane} \qquad (31)$$

Presumably excess phosphorus containing ligand suppresses reactions (30) and (31) by effectively competing with HCN for coordinations sites.

Alkali metal and tetraalkylammonium borohydrides have been employed as promoters presumably to reduce Ni(II) cyanide complexes as they are formed, thereby returning them to the pool of active nickel complexes [12, 13].

Lewis acids, e.g. BR_3 compounds and salts of various metals ($ZnCl_2$, $SnCl_2$, etc.) have also been employed as important cocatalysts for the reaction [11, 13, 15]. The mechanism by which these metal salts promote catalyst activity is not known. It has been demonstrated, however, that in the stoichiometric reaction of $Ni[P(OC_2H_5)_3]_4$ with HCN, the presence of $ZnCl_2$ results in the formation of a cationic pentacoordinated nickel hydride $[HNi[P(OC_2H_5)_3]_4]^+[ZnCl_2CN]^-$ rather than the expected $HNi[P(OC_2H_5)_3]_3CN$ compound [23a].

In the presence of Lewis acid cocatalysts the isomerizing properties of the catalytic system are greatly increased [20]. This is of interest in the case of the isomerization of 3-pentenitrile to 4-pentenenitrile which is subsequently reacted with HCN to form adiponitrile [51, 55, 56].

5. HYDROCYANATION OF ACTIVATED OLEFINS

Addition of HCN to olefins containing activating substituents, e.g. aryl, —COOR, —CN, COR, NO_2, commonly occurs in the presence of basic catalysts such as cyanide ion itself [32]:

$$CN^- + CH_2{=}CH{-}C{\equiv}N \longrightarrow [NC{-}CH_2{-}CH{=}C{=}N^\ominus] \longrightarrow$$

$$\xrightarrow{H^\oplus} NC{-}CH_2CH_2{-}CN \qquad (32)$$

Other catalysts have been found effective for hydrocyanations of this type. For example, it is reported that HCN addition to α,β-unsaturated ketones can be achieved with a high degree of stereospecificity by the use of HCN adducts of aluminum alkyls as catalysts [33].

It is not surprising to find that transition metal complexes are also effective catalysts (see Table 2).

Some of these complexes, e.g. $Co_2(CO)_8$ [7], $CoH[P(OPh)_3]_4$ [9], $Ni[P(OR)_3]_4$ [12], and $Pd[P(OR)_3]_4$ [17], have already been discussed in connection with hydrocyanation of non-activated olefins. Others, e.g. $[Ni(CN)_4]^{4-}$ [34], $Ni(CO)_4$ [35], and Cu_2Cl_2 [36, 37], apparently do not catalyze HCN addition to non-activated olefins.

Effort in this area has been limited mainly to conjugated diolefins. Esters of acrylic and methacrylic acids, acrylonitrile, butadiene, styrene, α-nitrostyrene, 1-vinylcyclohex-1-ene, and 4-methyl-3-penten-2-one are representative of the olefin type investigated. Conjugate 1,4-addition occurs preferentially with some of these olefin. This is the case of butadiene which gives 3-pentenenitrile as the major product in the presence of $Ni(CO)_4$ [35], $Ni[P(OR)_3]_4$ [12], $CoH[P(OPh_3]_4$ [9], and Cu_2Cl_2 [36, 37]:

$$CH_2{=}CH{-}CH{=}CH_2 + HCN \longrightarrow CH_3CH{=}CH{-}CH_2CN \qquad (33)$$

In the $Co_2(CO)_8$-catalyzed reaction of 1-vinyl-1-cyclohexene, however, 1,2-addition is favoured:

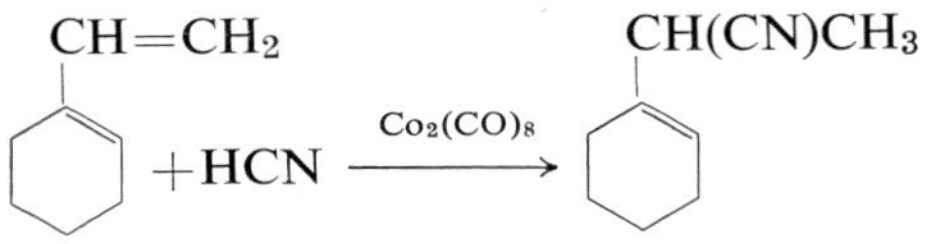

$$(34)$$

Table 2

HOMOGENEOUS HYDROCYANATION OF SOME ACTIVATED OLEFINS

Catalyst	Olefin	Product(s)	Ref.
Co₂(CO)₈	$C_6H_5CH=CH_2$	$C_6H_5CH(CN)CH_3$	[7]
	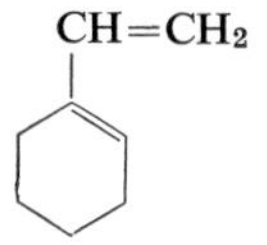 $CH=CH_2$ (1-vinylcyclohexene)	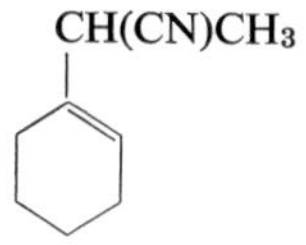 $CH(CN)CH_3$ (1-cyclohexenyl)	[7]
Hg[Co(CO)₄]₄	$CH_2=CHCH=CH_2$	$CH_2=CHCH(CN_3)CN$, $CH_3CH=CHCH_2CN$, $CH_2=CH(CH_2)_2CN$	[7]
CoH[P(OPh)₃]₄	$CH_2=CH-CH=CH_2$	$CH_3CH=CHCH_2CN$, $CH_2=CHCH(CH_3)CN$	[9]
Ni[P(OR)₃]₄, P(OR)₃, metal salt	$CH_2=CH-CH=CH_2$	$CH_3CH=CHCH_2CN$	[15]
	$C_6H_5CH=CH_2$	$C_6H_5CH(CN)CH_3$, $C_6H_5CH_2CH_2CN$	[14]
NiL₄(L=CO, P(OR₃)), BH₄⁻	$CH_2=CH-CH=CH_2$	$CH_3CH=CHCH_2CN$, $CH_2=CH(CH_2)_2CN$	[12, 13]
	$CH_2=C=CH_2$	$CH_2=CHCH_2CN$, $CH_2=C(CH_3)CN$, $CH_3CH=CHCN$	[12, 13]
K₄[Ni(CN)₄]	$CH_2=CH-COC_2H_5$ ($\overset{O}{\overset{\|}{}}$)	$NCCH_2CH_2COC_2H_5$ ($\overset{O}{\overset{\|}{}}$)	[34]
	$CH_2=CH-CN$	$NC(CH_2)_2CN$	[34]

Catalyst	Olefin	Product(s)	Ref.
	$(CH_3)_2C{=}CH{-}\overset{O}{\overset{\|}{C}}CH_3$	$(CH_3)_2C(CN)CH_2\overset{O}{\overset{\|}{C}}CH_3$	[34]
	$C_6H_5CH{=}CHNO_2$	$C_6H_5CH(CN)CH_2NO_2$	[34]
	$CH_2{=}CH{-}CH{=}CH_2$	mixture of monocyanobutenes and dicyanobutenes	[34]
$CaK_2[Ni(CN)_4]$	$CH_2{=}CH{-}CN$	$NC(CH_2)_2CN$	[34]
	$CH{=}C(CH_3)COOCH_3$	$NCCH_2CH(CH_3)COOCH_3$	[34]
$Na_4[Ni(CN)_4]$	$CH_2{=}C(CH_3)COOCH_3$	$CH_2(CN)CH(CH_3)COOCH_3$	[34]
	$CH_2{=}CH{-}COOC_2H_5$	$CH_2(CN)CH_2COOCH_3$	[34]
$Ni(CO)_4$	$CH_2{=}CH{-}CH{=}CH_2$	$CH_3CH{=}CHCH_2CN,\ CH_2{=}CH(CH_2)_2CN$	[35]
$Ni(CO)_4$, PPh_3	$CH_2{=}CH{-}CH{=}CH_2$	$CH_3CH{=}CHCH_2CN,\ CH_2{=}CH(CH_2)_2CN$	[35]
$Ni(CO)_4$, $AsPh_3$	$CH_2{=}CH{-}CH{=}CH_2$	$CH_3CH{=}CHCH_2CN,\ CH_2{=}CH(CH_2)_2CN$	[35]
$Pd[P(OR)_3]_4$, boron compound and metal salt	$C_6H_5CH{=}CH_2$	hydrocyanated styrene	[17]
Cu_2Cl_2	$CH_2{=}CH{-}CH{=}CH_2$	$CH_3CH{=}CHCH_2CN$	[37]
		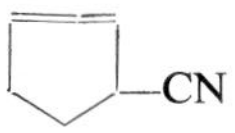	[36]

These variations may be dependent upon the nature of the catalyst. There are just too few examples to allow generalizations. The use of transition metal complexes as catalysts appears often to offer no real advantage over other procedures for HCN addition (Faster reaction rates have been claimed over some conventional methods [34]). Perhaps this accounts for the limited attention that this reaction has received. The hydrocyanation of butadiene is the only reaction fully studied. This reaction, which is the first step in the synthesis of adiponitrile, s however not treated in detail in this Review which is limited to mono-olefins.

6. RECENT RESULTS

Interest in the hydrocyanation of 3-pentenenitrile continues to run high. Several patents have issued recently disclosing processes by which 3-pentenenitrile is isomerized to the 4-isomer and subsequently reacted with HCN to form adiponitrile in the presence of ruthenium and iron or palladium and platinum complexes [47], [48] and Lewis acid cocatalysts.

The HCN addition reaction has been extended to olefins containing silyl substituents [49, 57]:

$$CH_2{=}CH{-}SiX_3 + HCN \xrightarrow{\text{PdL}_4} NCCH_2CH_2SiX_3$$

Employing tetrakis(triphenyphosphite) palladium (0) as catalyst, reasonable yields (30-84%) of cyanoethylsilanes were reported. Further, the β-cyanoethyl isomer predominated in the cases investigated.

Allyl triethoxysilane reacted similarly:

$$CH_2{=}CH{-}CH_2{-}Si(OC_2H_5)_3 + HCN \xrightarrow{\text{PdL}_4} NCCH_2CH_2CH_2Si(OC_2H_5)_3 +$$

$$+CH_3CH(CN)CH_2Si(OC_2H_5)_3$$

affording an equimolar mixture of cyanopropyltriethoxysilanes in about 50% yield. Concurrently, isomerization of the olefin to the propenyl isomer occurred. This latter olefin fails to react with HCN under these conditions.

The use of nichel zerovalent complexes with chelating diphosphines $Ph_2P{-}(CH_2)_n{-}PPh_2$ has been reported [51]. When $n = 4$ the selectivity to linear 3-and 4-pentenenitriles is very high (in the case of butadiene hydrocyanation).

With catalytic systems formed by $Ni(CO)_4$ + diphosphine [52] or $Fe(CO)_5$ + + diphosphine [53] the activity and selectivity are much lower. Interestingly $Ni[Ph_2P{-}(CH_2)_n{-}PPh_2]_2$ ($n = 2, 3, 4$) is a very effective catalyst of isomerisation also in the absence of a cocatalyst [51].

A recent report suggests that at least one major chemical firm is considering the commercial exploitation of the hydrocyanation of pentenenitrile for the synthesis of adiponitrile employing a zerovalent nickel catalyst system [50].

A process has been reported in which butadiene reacts with I_2 and copper (I) cyanide to produce the cuprous iodide complex of dehydroadiponitrile. Hydrolisis with aqueous HCN results in pure 1,4 dinitrile and regeneration of I_2 and copper cyanide [58].

7. CONCLUSION

In the previous chapters we have shown that the synthesis from olefins of a large variety of nitriles can easily be achieved using transition metal complexes as catalysts and this may be of interest in synthetic organic chemistry.

The activity and life of these catalysts (nickel (0) or palladium (0) complexes are rather good catalysts for this reaction and usually 200-300 cycles per mole of catalyst can be accomplished before deactivation), can be affected by factors such as excess of ligands or the use of activating cocatalysts (such as a Lewis acid or a reducing agent).

The use of activating cocatalysts is of interest because there is an industrial interest in the hydrocyanation of C_7—C_{12} olefin fractions and of pentenenitriles. High molecular weight olefins react with HCN under hydrocyanation conditions less easily than C_2—C_4 olefins. However the use of Lewis acids as cocatalysts has increased not only the rates of conversion in the case of high molecular weight olefins but has also allowed some control of the steric course of the reactions. This has been achieved by controlling the rate of double bond isomerization or the position of insertion into the double bond.

A great amount of work is still necessary to understand more clearly the mechanism of the catalytic reaction and to control and explain the action of the cocatalysts; the experimental investigation also needs to be extended to a larger number of olefins.

At the moment there is considerable industrial interest in the hydrocyanation of butadiene to give adiponitrile using a commercially economic process.

This subject, which will be treated in detail later in this series, is actually the driving force behind the development of the fundamental aspects of hydrocyanation reaction catalyzed by transition metal complexes.

In the near future this knowledge will certainly be of great help also to organic chemists interested in the synthesis of many different kind of nitriles and their derivatives.

8. REFERENCES

[1] D. T. MOWRY, "Chem. Revs.", *42*, 189, (1948).

[2] "Cyanides in Organic Chemistry, A Literature Review", Electrochemicals Department, E. I. du Pont de Nemours and Co., Wilmington, Delaware, (1962).

[3] See, for example, G. C. MONROE, Jr. and G. N. HAMMER, U. S. Patent 3,297,742; "Chem. Abstr.", *66*, 46117, (1967).

[4] See, for example, Brit. Patent 1,127,355; "Chem. Abstr.", *70*, 11157, (1969).

[5] See, for example, O. BAYER, "Angew. Chem.", *61*, 229, (1947).

[6] J. W. TETER, U. S. Patent 2,385,741; "Chem. Abstr." *40*, 590, (1946).

[7] *a*) P. ARTHUR, Jr., D. C. ENGLAND, B. C. PRATT and G. M. WHITMAN, "J. Amer. Chem. Soc.", *70*, 5364, (1954).

    b) P. ARTHUR, Jr. and B. C. PRATT, U. S. Patent 2,666,780; "Chem. Abstr.". *49*, 1776, (1955); U. S. Patent 2,666, 748; "Chem. Abstr.", *49*, 1774, (1955).

[8] E. A. RICK and E. S. BROWN, unpublished result.

[9] W. C. DRINKARD, Jr. and B. W. TAYLOR, Belg. Patent 723,382; Neth. Patent 68 15,812.

[10] W. C. DRINKARD, Jr., Ger. Patent 1,807,088; "Chem. Abstr.", *71*, 38383; (1969); Belg. Patent 723,381; Neth. Patent 68 15,746.

[11] W. C. DRINKARD, Jr. and R. J. KASSAL, Fr. Patent 1,529,134; "Chem. Abstr." *71*, 30092, (1969); Belg. Patent 700,420; Brit. Patent 1,146,330; Neth. Patent 67 08812.

[12] W. C. DRINKARD, Jr., U. S. Patent 3,496,218; Fr. Patent 1,544,658; Neth. Patent 67 06555; Brit. Patent 1,112,539; Belg. Patent 698,332.

[13] W. C. DRINKARD, Jr., Ger. Patent OLS 1,806,096; "Chem. Abstr." *71*, 30093. (1969); Neth. Patent 68 15560; Belg. Patent 723,126.

[14] W. C. DRINKARD, Jr. and R. V. LINDSEY, Jr. Brit. Patent 1,104,140; "Chem. Abstr" *68*, 77795, (1968); U. S. Patent 3,496,215; Belg. Patent 698,333: Neth. Patent 67 05556; U. S. Patent 3,496,217.

[15] YUAN-TSAN CHIA, W. C. DRINKARD, Jr. and E. N. SQUIRE, Brit. Patent 1,178,950; "Chem. Abstr.", *72*, 89831, (1970); Belg. Patent 720,659, Neth. Patent 68 12950.

[16] A. W. ANDERSON and M. O. UNGER, Neth. Patent 69 07449.

[17] W. C. DRINKARD, Jr. and R. V. LINDSEY, Jr., Ger. Patent 1,806,098; "Chem. Abstr.", *71*, 49343, (1969); Belg. Patent 723,128; Neth. Patent 68 15,487.

[18] E. A. RICK and E. S. BROWN, " J. Chem. Soc. D", 112, (1969).

[19] E. A. RICK and E. S. BROWN, *Abstracts Amer. Chem. Soc. Pet. Div. Chem. Reprints*, *14*, B29, (1969).

[20] W. C. DRINKARD, Jr. and B. W. TAYLOR, Ger. Patent 1,807,089; "Chem. Abstr.", *71*, 101,350, (1969); Belg. Patent 723,380; Neth. Patent 68 15,744 and 68 15,746.

[21] H. SINGER and G. WILKINSON, "J. Chem. Soc. A", 2516, (1968).

[22] L. BENZONI, "Chimica e Industria" (Milan), *50*, 1227, (1968).

[23] *a*) W. C. DRINKARD, Jr. and R. V. LINDSEY, Jr., Belg. Patent 723,386.

b) R. A. SCHUNN, "Inorg. Chem.", *9*, 394, (1970).

c) C. A. TOLMAN, "J. Amer. Chem. Soc.", *92*, 4217, (1970).

d) F. CARIATI, R. UGO and F. BONATI, "Inorg. Chem.", *5*, 1128, (1966).

[24] J. CHATT, R. S. COFFEY, A. GOUGH and D. T. THOMPSON, "J. Chem. Soc. A" 190, (1968).

[25] *a)* W. C. DRINKARD, Jr., D. R. EATON, J. P. JESSON, and R. V. LINDSEY, Jr., "Inorg. Chem.", *9*, 392, (1970).

b) R. A. SCHUNN, "Inorg. Chem.", *9*, 394, (1970).

[26] J. KWIATEK and J. K. SEYLER, "J. Organometal. Chem.", *3*, 433, (1965).

[27] *a)* M. D. JOHNSON, M. L. TOBE and LAI-YOONG WONG, "Chem. Comm.", 198, (1967).

b) M. D. JOHNSON, M. L. TOBE and LAI-YOONG WONG, "J. Chem. Soc. A", 491, (1967).

c) *ibid*, 923, (1968).

d) *ibid*, 929, (1969).

[28] R. B. KING, "Inorg. Chem.", *6*, 25, (1967).

[29] E. O. FISCHER, R. J. SCHNEIDER, "J. Organometal. Chem.", *12*, P 27, (1968).

[30] J. F. GUTTENBERGER, "Chem. Ber." *101*, 403, (1968).

[31] R. UGO, "Coord. Chem. Revs.", *3*, 319, (1968).

[32] P. KURTZ, "Annalen", *572*, 23, (1951).

[33] W. NAGATA and M. YOSHIOKA, "Yuki Gosei Kagaku Hyokai Shi", *26*, 2, (1968); "Chem. Abstr.", *68*, 77327, (1968).

[34] G. R. CORAOR and W. Z. HELDT, U. S. Patent 2,904, 581; "Chem. Abstr.", *54*, 4393; (1960); Brit. Patent 845,086; Can. Patent 628,973.

[35] P. ARTHUR, Jr. and B. C. PRATT, U. S. Patent 2,571,099; "Chem. Abstr.", *46*, 3068, (1952).

[36] W. A. SCHULZE and J. E. MAHAN, U. S. Patent 2,422,859; "Chem. Abstr.", *42*, 205, (1948).

[37] D. D. COFFMAN, L. F. SALISBURY, and W. D. SCOTT, U. S. Patent 2,509,859; "Chem. Abstr.", *44*, 8361, (1950).

[38] E. POLHAND, "Z. Anorg. Allg. Chem.", *201*, 282 (1931).

[39] E. POLHAND and W. HARLOS, "Z. Anorg. Allg. Chem.", *207*, 242, (1932).

[40] J. F. GUTTENBERGER, "Chem. Ber.", *101*, 403 (1968).

[41] J. COLMANN, "Accounts Chem. Res." *1*, 136, (1968).

[42] J. HALPERN, "Accounts Chem. Res.", *3*, 386, (1970).

[43] R. UGO, "Chimica e Industria" (Milan), *51*, 1319, (1969).

[44] *a)* J. C. KOTZ and D. G. PEDYOTTY, "Organomet. Chem. Rev. (A)", *4*, 479, (1969).
b) D. F. SHIVER, "Accounts Chem. Res.", *1*, 231, (1970).

[45] R. VAN DER LINDE and R. O. DE JONGH, "J. Chem. Soc. D", 563, (1971).

[46] Y. YAMAMOTO, H. YAMAZAKI, N. HAGIHARA, "J. Organomet. Chem"., *18*, 189, (1969).

[47] W. C. DRINKARD, Jr. and R. V. LINDSEY, Jr., U. S. Patent. 3,542,847 Nov. 24, 1970.

[48] W. C. DRINKARD, Jr. and R. V. LINDSEY, Jr., U.S. Patent 3,551,474 Dec. 29, 1970.

[49] E. S. BROWN, E. A. RICK and F. D. MENDICINO, *Abstracts*, *2nd North America Meeting of Catalysis Soc.* Houston - Feb. 22-24 p. 20 (1971).

[50] *Chemical Engineering News*, *April* 26, 1971, p. 30.

[51] *a*) P. ALBANESE, L. BENZONI, B. CORAIN and A. TURCO, Dom. It. Patent 13592 A/69;

   b) P. PASQUINO, L. BENZONI, G. CARNISIO and L. COLOMBO, Dom. It. Patent 25901 A/69.

[52] P. ALBANESE, L. BENZONI, G. CARNISIO and A. CRIVELLI, Dom. It. Patent 20837 A/69.

[53] P. ALBANESE, L. BENZONI and G. CARNISIO, Dom. It. Patent 20838 A/69.

[54] YUAN-TSAN CHIA, Neth. Patent 68 15,313.

[55] W. C. DRINKARD, Jr. and R. J. KASSAL, Belg. Patent 700,420; French Patent 1,529,134.

[56] YUAN-TSAN CHIA, W. C. DRINCKARD, Jr. and E. N. SQUIRE, Neth. Patent 68 12,976.

[57] E. S. BROWN, E. A. RICK and F. D. MENDICINO, " J. Organomet. Chem. ", *38*, 37, (1972).

[58] *Chemical Engineering News*, *April* 5, 1971, p. 31.

Chapter 3

Nickel Catalysed Syntheses
of Methyl-Substituted Cyclic Olefins,
an Example of Stepwise Carbon-Carbon
Bond Formation Promoted
by a Transition Metal Complex

P. HEIMBACH

Max-Planck-Institut für Kohlenforschung Mülheim, Ruhr (Germany)

based on experiments and ideas of
H. BUCHHOLZ, W. FLECK, P. HEIMBACH, D. HENNEBERG, H. HEY, H. LEHMKUHL, G. SCHOMBURG, H. SELBECK, F. THÖMEL and G. WILKE
Translated by W. V. Dahlhoff

1. INTRODUCTION

The cyclotrimerization of buta-1,3-diene to cyclododeca-1,5,9-triene was discovered by G. Wilke in 1956 during an investigation concerned with the effect of titanium and chromium containing Ziegler catalysts on 1,3-diolefins. In the presence of a $TiCl_4/(C_2H_5)_2AlCl$-catalyst butadiene reacts forming predominantly *trans,trans,cis*-cyclododeca-1,5,9-triene, whereas on using a catalyst prepared from chromul chloride and aluminium triethyl a mixture of 60% all-*trans*- and 40% *trans,trans,cis*-cyclododeca-1,5,9-triene is obtained [1].

As early as 1954 H. W. B. Reed had shown that butadiene can couple to give *cis,cis*-cycloocta-1,5-diene in the presence of a low valent nickel catalyst [2]. He used the so-called "Reppe catalysts" of the type $L_2Ni(CO)_2$ (L = phosphane or phosphite) [3] which have to be activated by preliminary treatment with acetylene. A thorough investigation of low valent nickel catalysts only became possible after Wilke et al. succeeded in preparing carbonyl free zerovalent nickel complexes [4]. This meant that "tailor-made" catalysts could be made with exact stoichiometric ratios of metal to ligand.

Catalytically active complexes of zerovalent nickel can easily be obtained by reduction of nickel (II) acetylacetonate with ethoxyaluminium diethyl or aluminium triethyl in the presence of the required ligands. Phosphanes and phosphites proved themselves to be especially effective as ligands, as have mono-, di- or tri-olefins e.g. butadiene, *cis,cis*-cycloocta-1,5-diene or cyclododeca-1,5,9--triene. Partial precipitation of elementary nickel results if the reduction is carried out with insufficient amounts of these ligands.

Figure 1 summarizes the ring syntheses which have been carried out in the presence of zerovalent nickel catalysts. These reactions have been extensively reviewed [5, 6, 7, 8]:

If butadiene is the only available ligand for the reduced nickel, then three of the four possible isomers of cyclododeca-1,5,9-triene are formed [9]. However,

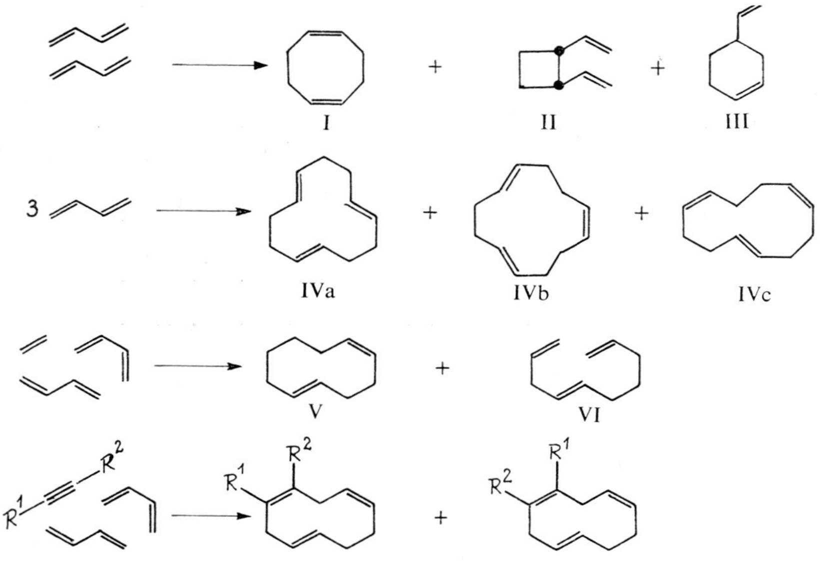

Figure 1. Syntheses of various ring systems promoted by nickel catalysts.

when one position of the nickel coordination sphere is blocked, total conversion of butadiene results in the formation of predominantly *cis,cis*-cycloocta-1,5-diene, whereas 40% *cis*-1,2-divinylcyclobutane is also formed at lower temperatures in cases where the conversion of the 1,3-diene does not exceed 85% [10]. *Cis,trans*-cyclodeca-1,5-diene is formed in high yield when butadiene and ethylene are reacted simultaneously [11]. Using dialkyl-, alkylaryl- and diarylsubstituted alkynes in place of ethylene results in the formation of 4,5-disubstituted *cis,cis-trans*-cyclodeca-1,4,7-trienes in equally high yields [12, 13, 14].

The intention of this article is to give a comprehensive account of the nickel catalyzed syntheses of mono and dimethyl derivatives of the above named 4-, 8-, 10- and 12-membered rings and their identification. The yields of some reactions promoted by a nickel triphenylphosphite catalyst are summarized in the following equations:

a) Cyclodimerization of *trans*-piperylene (see chapter 6.1.):

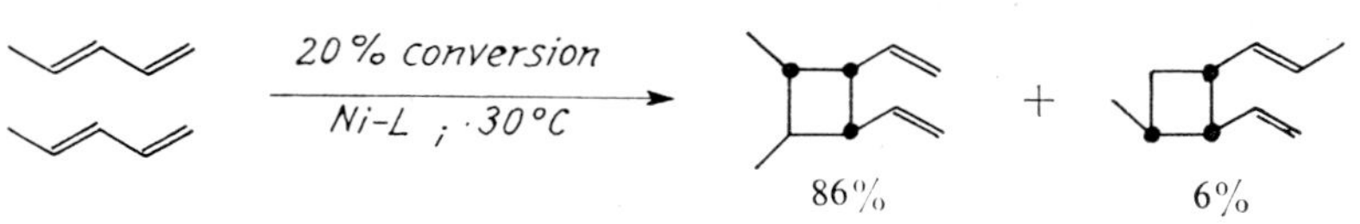

in this case $L = P(C_6H_{11})_3$

b) cross-co-cyclotrimerization (*) of butadiene, isoprene and ethylene (chapter 6.2.):

$$\xrightarrow[\text{[Ni-L]}]{40°C}$$

23% (¹)   +   41% (²)   +   37% (²)   +   22% (²)

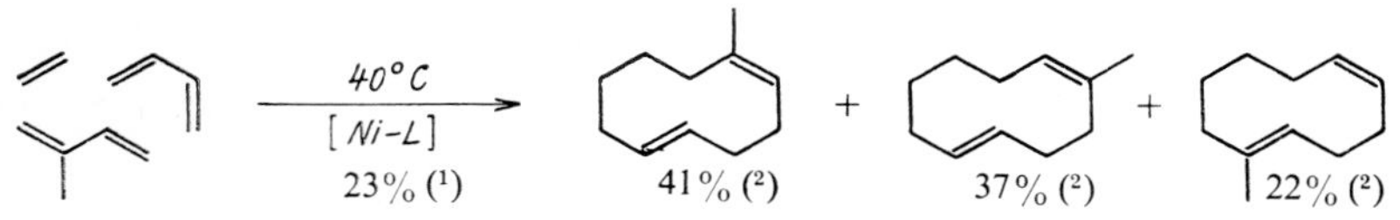

c) cross-cyclotrimerization of butadiene and isoprene (chapter 6.3.):

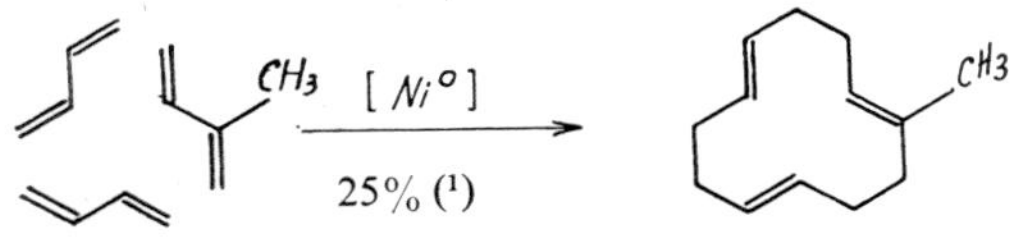

$$\xrightarrow[\text{25\% (¹)}]{\text{[Ni°]}}$$

d) co-cyclotrimerization of butadiene and propylene (chapter 6.2.):

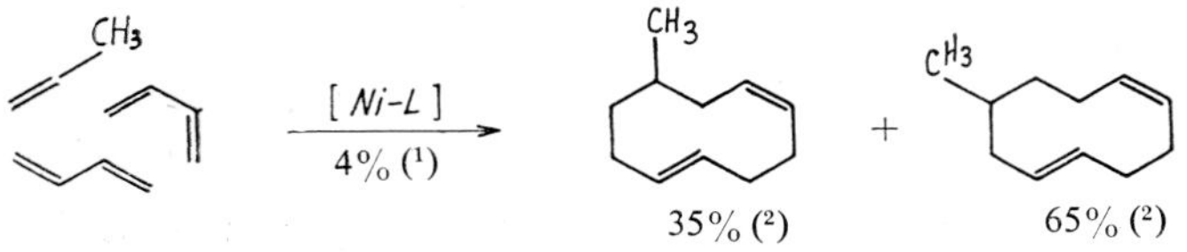

$$\xrightarrow[\text{4\% (¹)}]{\text{[Ni-L]}}$$

35% (²)   +   65% (²)

e) co-trimerization of isoprene and ethylene (chapter 6.2.):

$$\xrightarrow[\text{94\% (¹)}]{\text{[Ni-L]}}$$

1% (²)   +   43% (²)   +   3% (²)   +

28% (²)   +   9% (²)   +   10% (²)   +   6% (²)

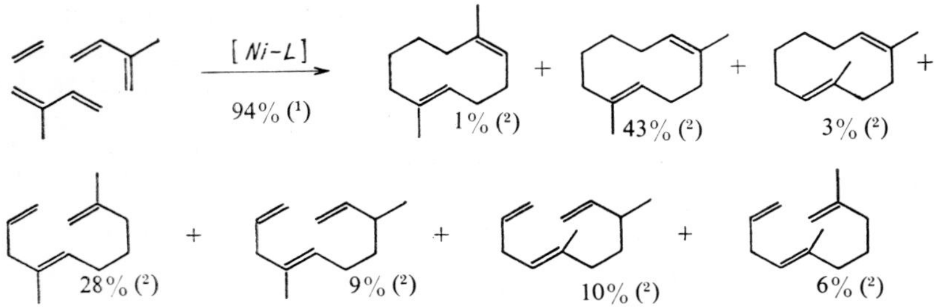

(*) In this review the following nomenclature will be adopted: *cross-dimerization, cross-trimerization or cross-oligomerization*: a di-, tri- or oligomerization of unsaturated compounds of the same type with different substituents or configurations, e.g. a dimerization of a butadiene molecule with a molecule of isoprene or a *trans*-with a *cis*-piperylene molecule. *Co-dimerization, co-trimerization or co-oligomerization*: a di-, tri- or oligomerization of unsaturated of different types, e.g. a trimerization of two molecules of a 1,3-diene with one molecule of ethylene or of one 1,3-diene with two alkynes. Consequently, a *cross-co-trimerization* is for example a trimerization of one molecule of ethylene with two different 1,3- dienes or of two different alkynes with a 1,3-diene molecule.

(¹) % proportion of the reaction mixture.

(²) Percentage distribution of the individual isomers.

f) co-cyclotrimerization of a mixture of *cis-* and *trans*-piperylene with ethylene (chapter 6.2.):

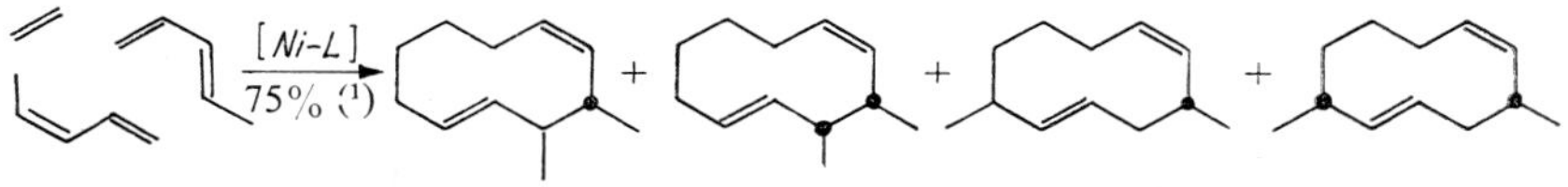

The experimental details of these nickel catalyzed reactions will be described in chapter 6. In addition in chapters 7. and 8. the mechanism and some general aspects of carbon-carbon bond formation promoted by transition metals will be discussed. The reason we took this task upon us is due to several factors:

1. We had investigated the preparative extension of the ring syntheses.

2. We hoped to obtain further information about the mechanism by using the methyl group as a disturbing influence in the 1,3-dienes, in the monoolefins and in the intermediate complexes.

3. We wanted to investigate which stereochemical factors are of importance in catalytic carbon-carbon coupling reactions.

Two reactions contributed substantially to the elucidation of the structures and configurations of the mono-and dimethylsubstituted ring compounds:

1. The thermal and "catalytic" Cope-Rearrangement of 1,5-dienes.

2. The methylene insertion reaction (MIR).

In addition the effective and reproducible separation and identification of the various isomers using gas-chromatography is a prerequisite of vital importance. The elucidation of the configurations of the catalytically formed products and their thermal rearrangements are significant aspects in the understanding of carbon-carbon bond formation promoted by homogeneous low valent nickel catalysts and therefore will be discussed in detail in the following chapters.

2. COPE REARRANGEMENT OF 1,5-DIENES

2.1. Cope Rearrangement of open-chain 1,5-dienes

The first examples of a thermal rearrangement in substituted open-chain 1,5-dienes were observed by A. C. Cope et al. [15]:

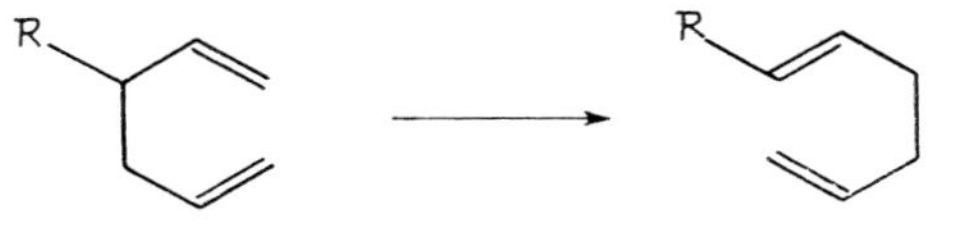

(1)

Investigations by W. v. E. Doering and W. R. Roth [16] led to specific ideas concerning the stereochemical pathway involved in this reaction. In a specific example they were able to show conclusively that in the case of open-chain 1,5-dienes the reaction proceeds via a four-centre arrangement ("chair" form) rather then via the alternative six-centre arrangement ("boat" form).

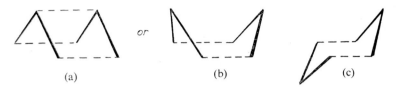

(a)   or   (b)   (c)

However, as is apparent in (b), only four carbon atoms may interact in the "boat" form.

They found that e.g. racemic-3,4-dimethyl-hexa-1,5-diene only gives 10% *cis,cis*- and 90% *trans,trans*-octa-2,6-diene on thermal rearrangement:

$$\text{(structure)} \longrightarrow \text{trans,trans} + \text{cis,cis} \tag{2}$$

A so called six-centred transition state would lead to the *cis,trans*-octa-2,6-diene:

$$\text{(structure)} \longrightarrow \text{trans,cis} \tag{3}$$

Whether the "boat" or the "chair" conformation is preferred in cyclic 1,5-dienes depends on the size of the ring system under investigation. The thermal rearrangements of *cis*-1,2-divinylcyclobutanes and *cis,trans*-cyclodeca-1,5-dienes were of great importance in the elucidation of the configurations and structures of the catalytically formed dimethylsubstituted *cis,cis*-cycloocta-1,5-dienes and the mono-and dimethylsubstituted ten membered rings.

2.2. Cope Rearrangement of cis-1,2-divinylcyclobutanes

As shown in the thermally initiated valence isomerization of cis-1,2-divinyl-cyclobutane to cis,cis-cycloocta-1,5-diene, discovered by E. Vogel [17], the Cope Rearrangement may also proceed via a so called six-centre mechanism under specific circumstances. A four-centre transition state would lead to the highly strained cis,trans-cycloocta-1,5-diene. However, we also found these products of a four-centre mechanism in rearranging certain methylsubstituted cis-1,2--divinylcyclobutanes, in which the "boat" like transition state is less favoured, perhaps because of steric hindrance. The four membered rings VIIa and VIIb are rearranged at 95 °C in 30% and 7% yields, respectively, to the corresponding cis,trans-cycloocta-1,5-dienes [18].

$$\text{(4)}$$

IXb 30%   VIIa   95°C   VIIa   VIIIc 70%

$$\text{(5)}$$

IXa 7%   VIIb   VIIb   VIIIb 93%

The trans-dimethyl-cis,trans-cycloocta-1,5-diene IXa and the isomer IXb were characterized as the phenylazide- and cyclopentadiene-adducts. IXb was isolated in a pure form and characterized by IR- and ^{1}H-NMR spectroscopy. The formation of ca. 3% of cis,trans-cycloocta-1,5-diene in the thermal rearrangement of cis-1,2-divinyl-cyclobutane had already been assumed by G. S. Hammond and DeBoer [19].

We were able to show that the "catalytic" Cope Rearrangement promoted by nickel ligand catalysts is more stereospecific than the thermal Cope Rearrangement [18, 20, 21] (see chapter 6.1., Figure 17).

2.3. Cope Rearrangement of cis,trans-cyclodeca-1,5-dienes

The thermally labile cyclodeca-1,5-dienes rearrange at 120-150 °C giving 1,2-divinylcyclohexanes [22]. We found that, like the unsubstituted cis,trans-isomer [23], the mono- and dimethyl-substituted cis-trans-cyclodeca-1,5-dienes also form the corresponding cis-1,2-divinylcyclohexanes via a four-centre mechanism at temperatures up to 180 °C [24]. Equilibria between certain substituted cis-trans-cyclodeca-1,5-dienes and the corresponding six-membered rings are only achieved after prolonged heating at 180 °C. In two recent theses [25, 26] these equilibria were investigated. The rearrangement of cis,trans-cyclodeca-

1,5-diene to *cis*-1,2-divinyl-cyclohexane has an activation-energy of ca. 27 kcal/mol, the retro Cope Rearrangement having a value of ca. 35-37 kcal/mol [25].

If the rearrangement is carried out below 150 °C it can be concluded from the configurations of the resulting six-membered rings at which position the medium sized ring was substituted (see later, equation 12). On the other hand, certain ten-membered rings may be synthesized from the corresponding cyclohexane derivatives.

In Figure 2 is a general scheme showing the rearrangements and possible equilibria of substituted *cis,trans*-cyclodeca-1,5-dienes and *cis*-1,2-divinylcyclohexanes. This scheme is only valid for the Cope and retro Cope Rearrangements, which occur via the "chair" form. In the cyclodeca-1,5-diene-derivatives of type A (see Figure 2) the substituents (3′), (4′) and (7′) to (10′) are in axial positions. In the rings of type B these substituents are equatorial. The same applies to the six-membered rings. The substituents (3′) and (4′) on the double bonds in the cyclohexanes of type A are *cis*, those in type B are *trans* to the six-membered ring. Four different equilibria between conformers have to be considered in this scheme:

$$A^1 \rightleftharpoons B^1 \qquad\qquad A^4 \rightleftharpoons B^4$$
$$A^2 \rightleftharpoons B^3 \qquad\qquad A^3 \rightleftharpoons B^2$$

As mentioned earlier, evidence for the back reactions:

$$A^2 \longrightarrow A^1 \qquad\qquad B^2 \longrightarrow B^1$$
$$A^3 \longrightarrow A^4 \qquad\qquad B^3 \longrightarrow B^4$$

can only be found by prolonged heating to 180 °C of those cyclodeca-1,5-dienes which have methylsubstituted double bonds. This observation was of interest concerning the intended syntheses of certain *cis,trans*-cyclodeca-1,5-diene derivatives.

The three cross-co-trimers Xa, Xb and Xd are formed from butadiene, isoprene and ethylene in a reaction promoted by a nickel ligand catalyst (see chapter 3.2.). They rearrange on heating to XIa and XIb. The cyclodeca-1,5-diene Xc, which is not formed catalytically, appears, together with Xa, from XIb by heating at 180 °C for 15 hours:

$A^1 \rightleftharpoons B^1$; $1 = CH_3$
Xb

$A^2 \rightleftharpoons B^2$; $1 = CH_3$
$B^2 \rightleftharpoons A^3$ bzw.
$6 = CH_3$

(6)

XIa

$A^1 \rightleftharpoons B^1$; $6 = CH_3$
Xd

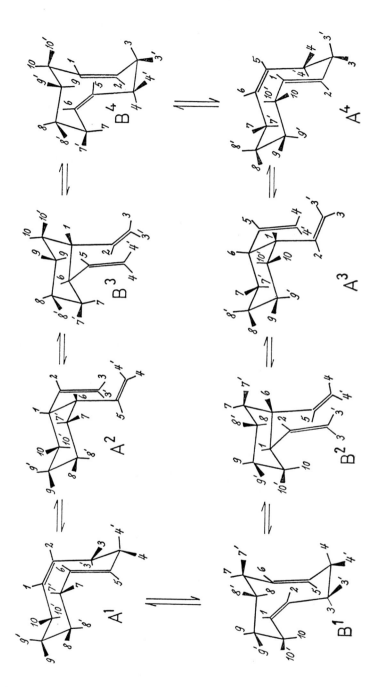

Figure 2. General scheme for the rearrangement of *cis,trans*-cyclodeca-1,5-dienes and the equilibria between six- and ten-membered rings via a « chair »-form in the transition state [24].

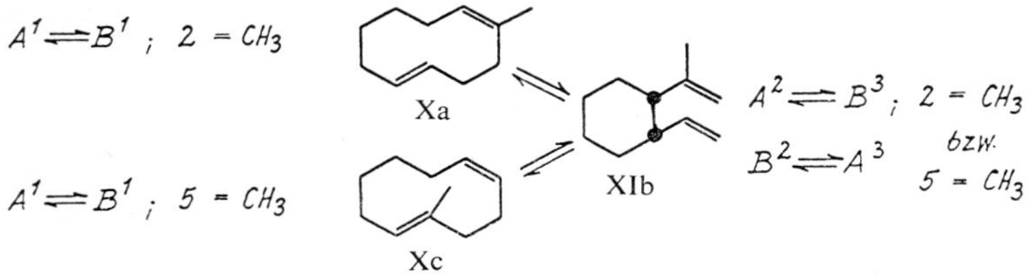

$A^1 \Longleftrightarrow B^1$; $2 = CH_3$

Xa

$A^2 \Longleftrightarrow B^3$; $2 = CH_3$
bzw.
$B^2 \Longleftrightarrow A^3$ $5 = CH_3$

(7)

XIb

$A^1 \Longleftrightarrow B^1$; $5 = CH_3$

Xc

Xb can be isolated from the mixture of Xa, Xb and Xd by the following procedure: after heating this mixture at 150 °C for three hours, less than 1% of Xa and Xd can be detected; yet 18% of Xb is still present. Xb can then be isolated by preparative gas-chromatography.

We found very similar results in the case of the dimethylsubstituted cyclodeca-1,5-dienes XIIa, XIIb and XIId, which were synthesized from two molecules of isoprene and ethylene (see chapter 6.2.):

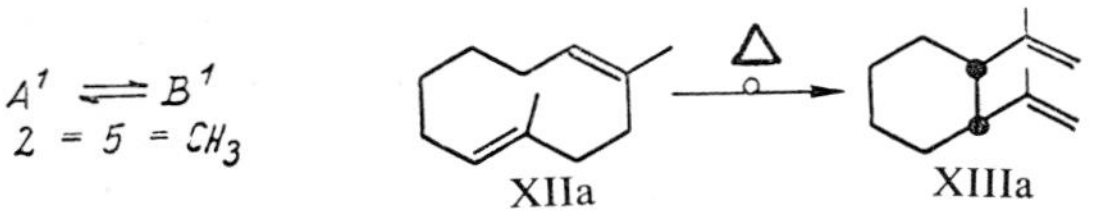

$A^1 \Longleftrightarrow B^1$
$2 = 5 = CH_3$

XIIa → XIIIa

(8)

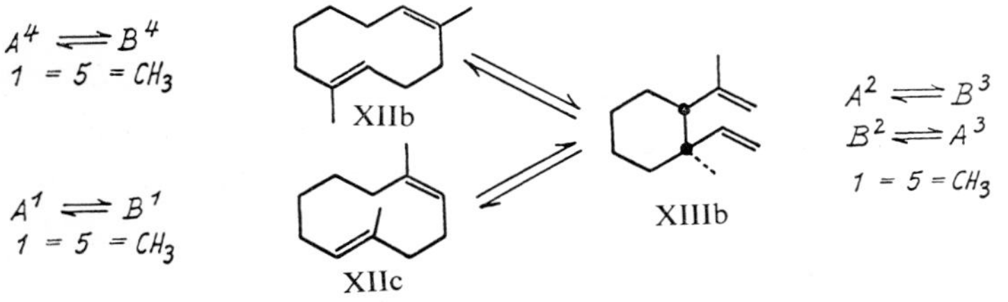

$A^4 \Longleftrightarrow B^4$
$1 = 5 = CH_3$

XIIb

$A^1 \Longleftrightarrow B^1$
$1 = 5 = CH_3$

XIIc

XIIIb

$A^2 \Longleftrightarrow B^3$
$B^2 \Longleftrightarrow A^3$
$1 = 5 = CH_3$

(9)

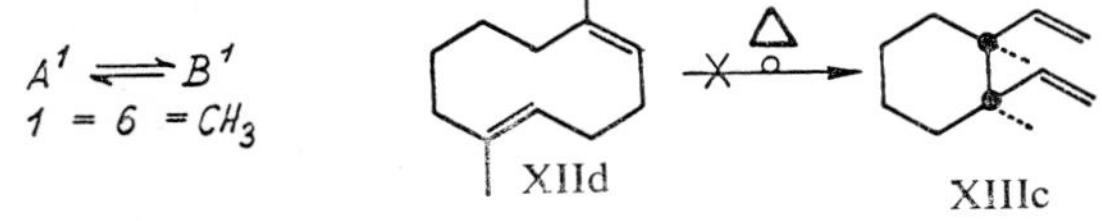

$A^1 \Longleftrightarrow B^1$
$1 = 6 = CH_3$

XIId

XIIIc

(10)

However, XIId is stable at 150 °C and also to prolonged heating at 180 °C. After heating XIIIb to 180 °C for fifteen hours 13% of XIIb and 7% of XIIc could be identified (for further evidence of the configurations of XIIa, XIIb, XIIc and XIId see chapter 4.2.). There is good experimental evidence that all isomers, in which the methyl group is not attached to a double bond, rearrange predominantly via those conformations in which the methyl groups are in equatorial positions. By cross-co-oligomerization of butadiene, piperylene and ethylene the isomers XIVb, XIVc and XIVd are formed (see chapter 6.2.). These isomers and also XIVa, which is not formed catalytically, rearrange thermally according to the following equations:

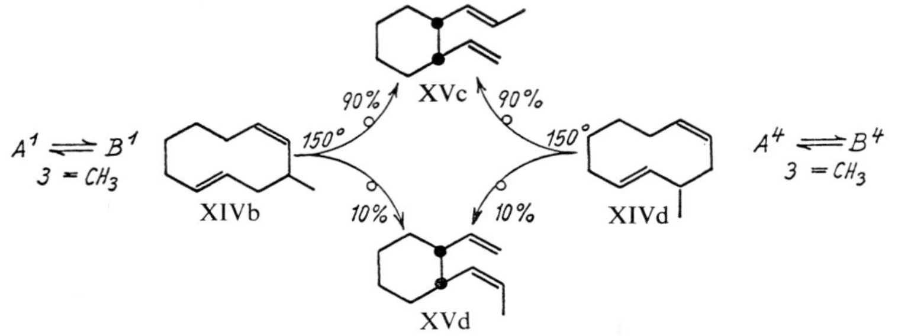

$$A^1 \rightleftharpoons B^1 \atop 3 = CH_3 \qquad A^4 \rightleftharpoons B^4 \atop 3 = CH_3 \qquad (11)$$

$$A^1 \rightleftharpoons B^1 \atop 7 = CH_3 \qquad A^4 \rightleftharpoons B^4 \atop 7 = CH_3 \qquad (12)$$

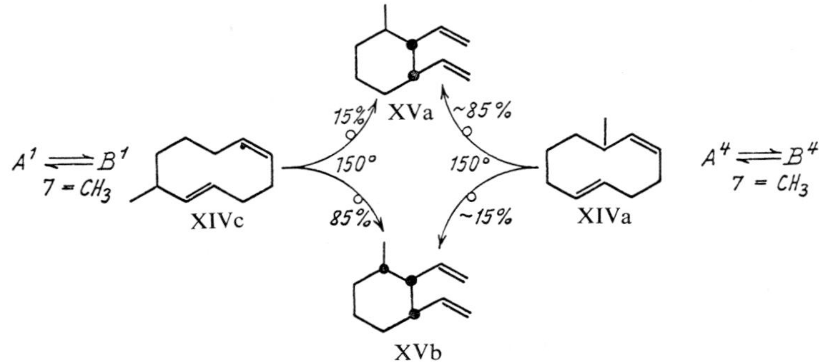

From the thermal valence-isomerization of the catalytically formed XIVc to 85% XVb (*) and 15% XVa, we were able to determine the position of the methyl group in the ten-membered ring. However, we could only show indirectly, that XIVa rearranges as predicted in equation 12. XIVa could only be synthesized by MIR together with all other monomethyl derivatives of cis,trans-cyclodeca-1,5-diene (see chapter 3.2). Heating the mixture, in which XIVa and XIVc are present in a ratio 1 : 1, to 150 °C for several hours leads to all corresponding methyl-cis-1,2-divinylcyclohexanes. Because the isomers XVa and XVb are present in the ratio 1 : 1, one can conclude that XIVa forms predominantly XVa.

The same considerations adopted in the case of the isomers XIVa and XIVc are valid for the isomers XVIa and XVIb, which are formed catalytically from two molecules of butadiene and propylene:

(*) The configuration of XVb was elucidated as follows:

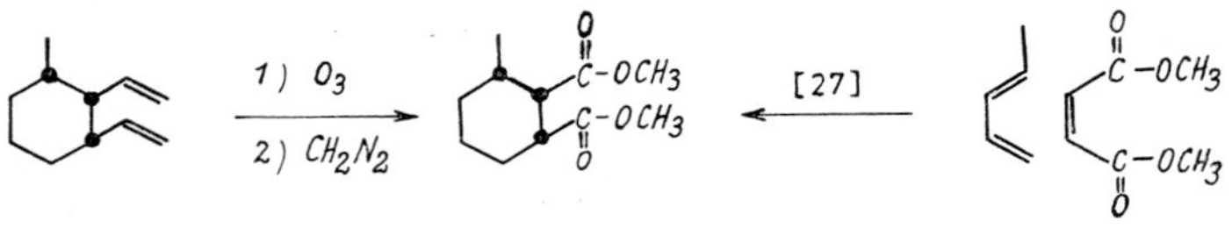

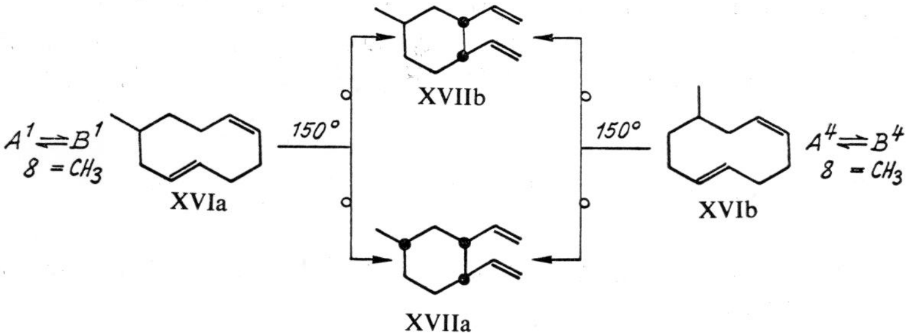

$$A^1 \rightleftharpoons B^1 \quad \xrightarrow{150°} \quad \text{XVIIb} \quad \xleftarrow{150°} \quad A^4 \rightleftharpoons B^4 \quad (13)$$
$$8 = CH_3 \qquad\qquad\qquad\qquad\qquad\qquad\qquad\qquad 8 = CH_3$$
$$\text{XVIa} \qquad\qquad\qquad\qquad\qquad\qquad\qquad\qquad \text{XVIb}$$
$$\text{XVIIa}$$

The ten-membered rings XVIa and XVIb could be isolated in a ratio 72 : 28. After thermal rearrangement the corresponding six-membered rings XVIIb and XVIIa were present in a ratio 77 : 23 as shown by gas chromatographic analysis (for further evidence for the structures of XVIa and XVIb see chapter 6.2. equation 39).

The configurations of the isomers XVIIIa to XVIIId, which are formed from piperylene and ethylene, could be identified by elucidation of the structures and configurations of the thermally formed six-membered rings XIXa to XIXd from their IR- and ^{1}H-NMR spectra:

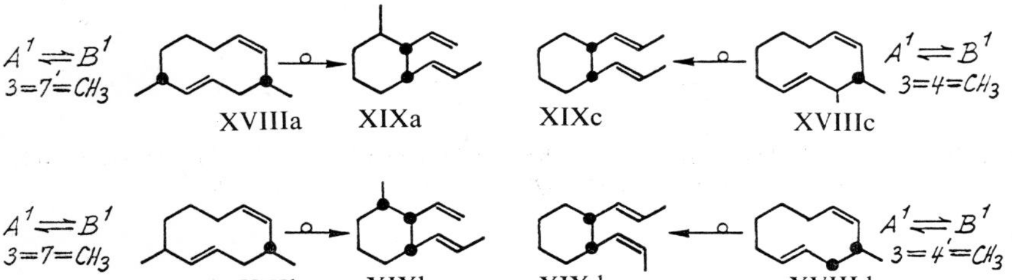

$$A^1 \rightleftharpoons B^1 \qquad\qquad\qquad\qquad\qquad\qquad A^1 \rightleftharpoons B^1 \quad (14)$$
$$3=7=CH_3 \qquad\qquad\qquad\qquad\qquad\qquad\qquad 3=4=CH_3$$
$$\text{XVIIIa} \quad \text{XIXa} \qquad \text{XIXc} \qquad \text{XVIIIc}$$

$$A^1 \rightleftharpoons B^1 \qquad\qquad\qquad\qquad\qquad\qquad A^1 \rightleftharpoons B^1 \quad (15)$$
$$3=7=CH_3 \qquad\qquad\qquad\qquad\qquad\qquad\qquad 3=4=CH_3$$
$$\text{XVIIIb} \quad \text{XIXb} \qquad \text{XIXd} \qquad \text{XVIIId}$$

In principle, the isomers XIXa and XIXb may also arise from the following rings:

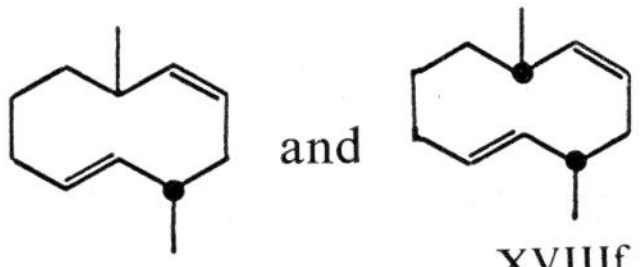

$$\text{and}$$
$$\text{XVIIIe} \qquad\qquad \text{XVIIIf}$$

but we assume that these rings are not formed in the catalytic reaction (see chapter 6.2.).

During a qualitative investigation of the rearrangement rates of those *cis*, *trans*-cyclodeca-1,5-dienes, in which one or two hydrogens on a double bond are substituted by one or two methyl groups, we observed the following rule: the smaller the difference in the number of alkyl substituents on the double bonds in the ten- and the six-membered rings (see Figure 3) the faster is the thermal Cope-Rearrangement:

$$XIIa > XIIb >>> XIId \quad and \quad Xa > Xd > Xb$$

In these isomers the energy content of the products seems also to influence the activation energies.

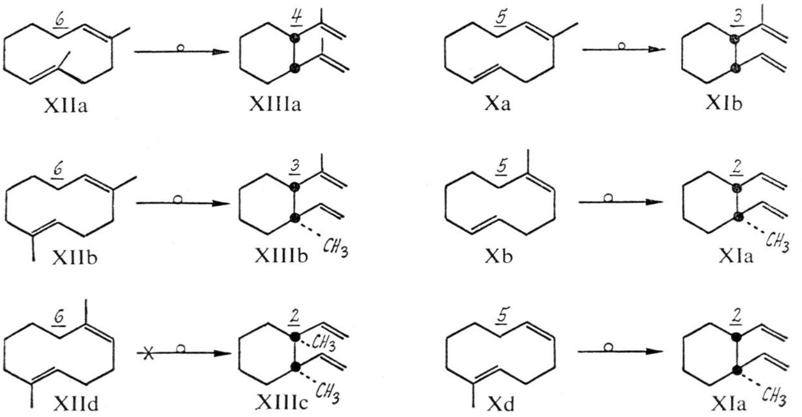

Figure 3. The number of alkyl substituents on the double bonds (underlined) in *cis,trans*-cyclodeca-1,5-diene and *cis*-1,2-divinylcyclohexane derivatives [24].

3. PREPARATION OF MONOMETHYL DERIVATIVES BY METHYLENE INSERTION REACTION (MIR)

Diazomethane decays on irradiation into nitrogen and highly reactive methylene. It has been reported by W. v. E. Doering et al. [28] that, in the presence of hydrocarbons, the methylene is nearly statistically inserted into all the carbon-hydrogen bonds of the substrate. Richardson, Dvoretzky et al. [29, 30] succeeded in identifying a great number of homologues of saturated and unsaturated hydrocarbons. Their "study surveys the reactivity of methylene from photolysis of diazomethane with representative alkanes, cycloalkanes, and olefins. With each of these hydrocarbon types, the insertion of methylene into the carbon-hydrogen bonds of the parent molecule allows the predictable synthesis of isomeric next-higher homologs, in spite of certain complications in the case of olefins" [30].

When double bonds are present in the molecules, cyclopropane derivatives are formed to a greater extent. *There is no alteration in the configurations of the original double bonds on forming the three membered rings and the homologues.* This is the decisive advantage of this reaction:

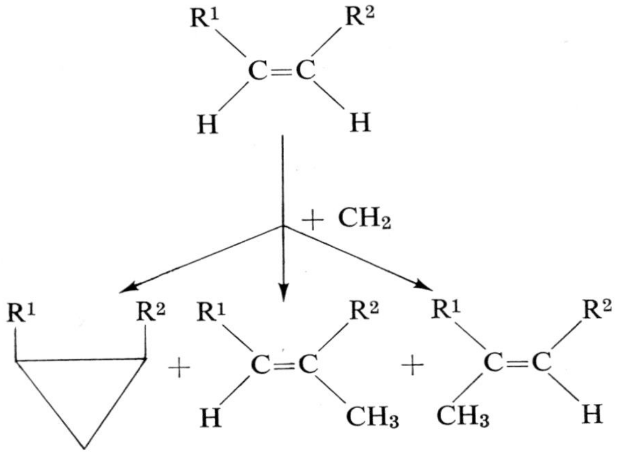

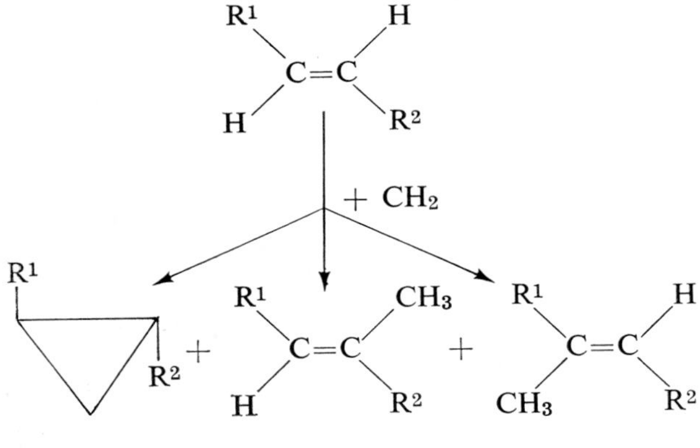

By using MIR one can easily determine the configurations of tri- and tetra-substituted olefins with one or two methyl groups or one ethyl group on the double bond, e.g.

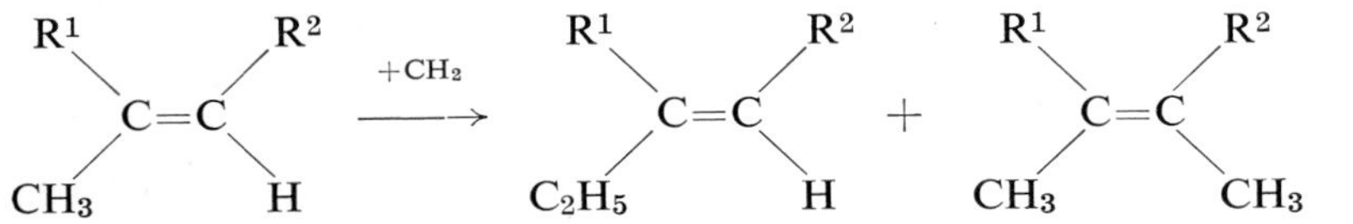

$$(17)$$

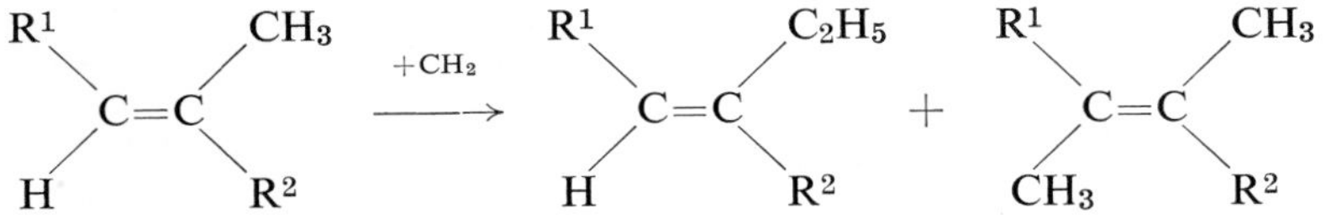

$$(18)$$

This method was used and expanded by G. Schomburg [31, 32] and other authors [33, 34]. G. Schomburg in particular tried to find general rules concerning the retention behaviour of specifically substituted molecules [35]. The information obtained by Schomburg is helpful in identifying the separate isomers of a MIR mixture.

The diazomethane is used only in very small concentrations (1-3 mole percent of the substrate) to avoid further reaction of the products with methylene. The solution of the diazomethane in the substrate is irradiated for ca. twenty minutes with a mercury-vapour lamp until the yellow colour of the diazomethane disappears. When a solution of diazomethane in olefins is heated in the presence of copper powder, only the cyclopropane derivatives are formed [37, 36]. Thus the cyclopropanes formed by MIR's can easily be identified.

One of the aspects of this review is to demonstrate how useful the preparation of mono- and dimethylsubstituted ring-olefins by MIR's is. At the same time I will try to show that a successful identification of all isomers is very often only possible together with information from catalytic reactions and thermal rearrangements.

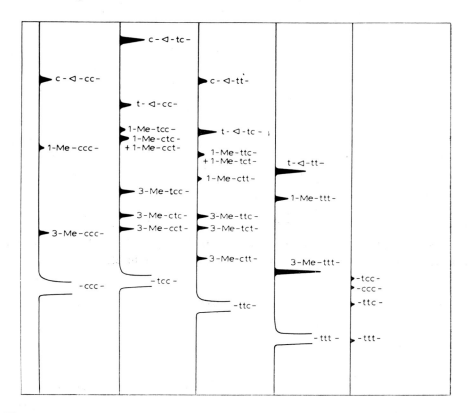

Figure 4. MIR mixtures of all possible cyclododeca-1,5,9-trienes. t, ▷–t, t = monocyclopropane derivative of the ttt-isomer [32].

3.1. Preparation and identification of all monomethyl derivatives of ttt-, ttc-, tcc- and ccc-cyclododeca-1,5,9-triene

Using transition metal catalysts only three of the four possible isomers of cyclododeca-1,5,9-triene with the configurations IVa, IVb and IVc are formed from butadiene. The four possible isomers are: ttt, ttc, tcc, ccc. The all-*cis*-isomer IVd was synthesized by Untch and Martin [38]. Figure 4 shows the MIR-mixtures of all isomers of cyclododeca-1,5,9-triene. G. Schomburg was able to identify all the individual isomers, by the statistical weight in which they are formed, and by considering the rule, that "small structural variations correspond to definite differences of retention data" [32]. These influences are known from a series of open-chain and cyclic olefins [31, 35].

3.2. Preparation and identification of all monomethyl derivatives of *cis,trans*-cyclodeca-1,5-diene, *trans*-deca-1,4,9-triene and *cis*-1,2-divinylcyclohexane

Using MIR we synthesized all monomethyl-derivatives of *cis,trans*-cyclo-deca-1,5-diene, *trans*-deca-1,4,9-triene and *cis*-1,2-divinylcyclohexane [24, 39]. By reacting these olefins with diazomethane in the presence of copper powder at higher temperatures, we obtained all the cyclopropanes, which are also formed in the MIR's. All the prepared monomethyl derivatives are shown in Figure 5, 6 and 7. The numbers on the arrows in the figures give the statistical weight, in which the isomers are formed in the MIR's. In addition Figures 5 and 6 portray from which 1,3-dienes and mono-olefins the individual isomers may be formed by the catalyst. Figure 7 also shows from which methylsubstituted *cis,trans*-cyclodeca-1,5-dienes the six-membered rings arise. Some interesting information was obtained by thermal rearrangement of the MIR-mixture of *cis,trans*-cyclodeca-1,5-diene (see chapter 2.3. equations 11 to 15).

Success in using the combination of MIR and catalytic experiments requires that the following prerequisites are fulfilled:

1. The individual isomers have to be identified as fairly well separated peaks using high resolution capillary columns.

2. The retention indices of the separated isomers must be determinable with high precision and accuracy.

3. Using columns with liquid phases of varying polarity, the assignments must be reproducible.

The following gas-chromatographic system fulfilled these prerequisites:

The gas-chromatograph was equipped with a glass capillary column with polypropyleneglycol (PPG) or a steel capillary with methylsilicon oil (CD 200). The temperature of the column was 80-120 °C. The inlet tube had only a temperature of 120 °C, during analysis of the thermally labile ten-membered rings.

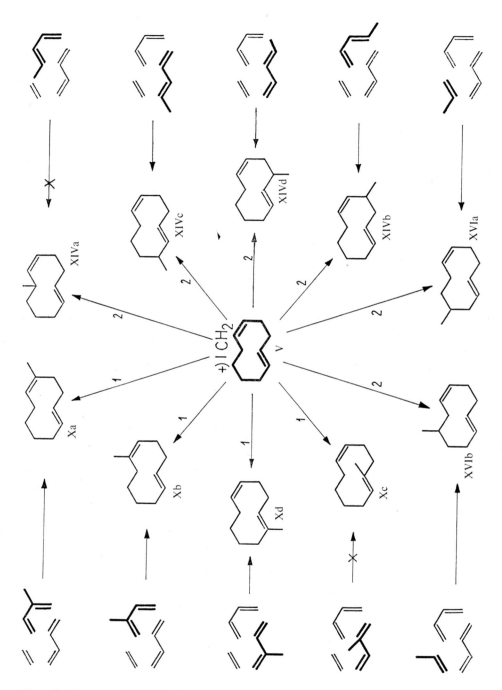

Figure 5. All monomethyl derivatives of *cis,trans*-cyclodeca-1,5-diene prepared by MIR. In addition the possible catalytic synthesis is shown [24].

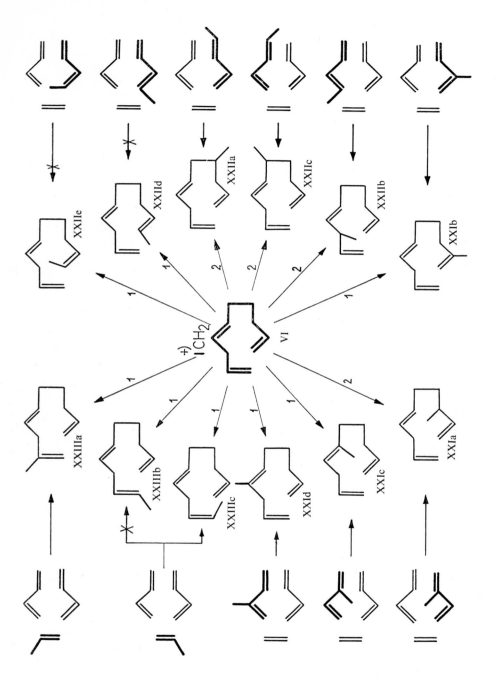

Figure 6. All monomethyl derivatives of *trans*-deca-1,4,9-triene prepared by MIR. In addition the possible catalytic synthesis is shown [24].

98

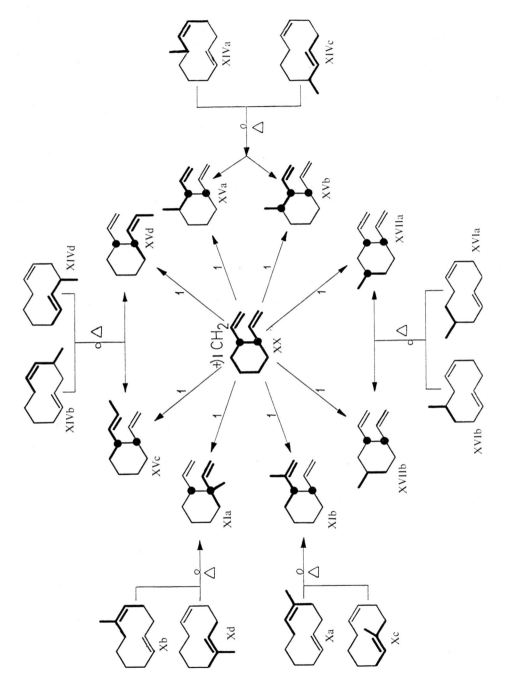

Figure 7. All monomethyl derivatives of *cis*-1,2-divinylcyclohexanes prepared by MIR. In addition the cyclodeca-1,5-dienes from which the isomers may be formed are shown [24].

The retention data were standardized with help of the Kovats retention index system, which correlates to the retention data of unbranched paraffins (e.g. n-octane = 800; n-dodecane = 1200) by logarithmic interpolation [40].

Table 1

KOVATS INDICES OF METHYL-*cis*-1,2-DIVINYLCYCLOHEXANES PREPARED BY MIR AND COPE REARRANGEMENT OF THE CORRESPONDING *cis,trans*-CYCLODECA-1,5-DIENES. (METHYL-SILICON OIL DC 200, 100 °C). [24].

Com-pound	Average-value of Kovats indices	MIR V	△	Monomethyl-derivatives of *cis*-1,2-divinylcyclohexane prep. by Cope Rearrangement of:					
				XIVa to XIVd	△	Xa, Xb, Xd	△	XVIa + XVIb	△
XVIIa	1047.8	1047.6	— 0.2					1047.9	+ 0.1
XVIIb	1047.8	1047.6	— 0.2					1047.9	+ 0.1
XVa	1049.9	1050.0	+ 0.1	1049.9	± 0.0				
XIa	1055.5	1055.5	± 0.0			1055.6	+ 0.1		
XVb	1056.7	1056.8	+ 0.1	1056.6	— 0.1				
XVc	1092.6	1092.6	± 0.0	1092.7	+ 0.1				
XIb	1097.1	1096.9	— 0.2			1097.2	+ 0.1		
XVd	1098.5	1098.6	+ 0.1	1098.4	— 0.1				

Table 2

KOVATS INDICES OF METHYL-*cis*-1,2-DIVINYLCYCLOHEXANES PREPARED BY MIR AND COPE-REARRANGEMENT OF THE CORRESPONDING *cis,trans*-CYCLODECA-1,5-DIENES. (POLY-PROPYLENGLYCOL PPG, 100 °C).

Com-pound	Average value of Kovats indices	MIR V	△	Monomethyl-derivatives of *cis*-1,2-divinyl-cyclohexane prep. by Cope Rearrangement of:					
				XIVa to XIVd	△	Xa, Xb, Xd	△	XVIa + XVIb	△
XVIIa	1102.3	1102.1	— 0.2					1102.6(*)	+ 0.3
XVIIb	1102.3	1102.1	— 0.2					1102.6(*)	+ 0.3
XVa	1102.2	1102.1	— 0.1	1102.3	+ 0.1				
XIa	1109.2	1109.1	— 0.1			1109.3	+ 0.1		
XVb	1112.3	1112.1	— 0.2	1112.5	+ 0.2				
XVc	7148.0	1147.8	— 0.2	1148.2	+ 0.2				
XVd	1155.3	1155.1	— 0.2	1155.4	+ 0.1				
XIb	1158.1	1157.9	— 0.2			1158.3	+ 0.2		

(*) Separated with squalane as liquid phase [24].

Determining the values of the retention data for the various isomers, we found deviations from the average value of 0,1 to 0,3 index units. In principle one has to consider, that the concentrations of the isomers (especially in the catalytically formed reaction mixtures) vary considerably and that the columns are sometimes overloaded for the species present in higher concentrations. Even under these circumstances excellent reproducibility was obtained. As an example the retention data of all methyl-*cis*-1,2-divinylcyclohexanes are given in tables 1 and 2. All index data relating to the co-oligomerization are listed in [24] and will be published elsewhere [39].

4. PREPARATION OF DIMETHYL DERIVATIVES BY MIR

In order to synthesize and identify dimethyl substituted cyclic olefins in a similar way to the monomethyl derivatives, it is necessary to isolate the mono-methyl substituted ring olefins in a pure form. 1-methyl-*cis,cis*-cycloocta-1,5-diene and 3-methyl-*cis,cis*-cycloocta-1,5-diene can easily be purified by fractionation. These isomers are formed in good yields (based on the converted substituted 1,3-diene) by the catalytic cross-cyclodimerization of butadiene with isoprene or piperylene [6, 7].

4.1. Preparation and identification of all dimethyl *cis,cis*-cycloocta-1,5-dienes

Besides the two ethyl- and four possible cyclopropyl-derivatives one obtains all dimethyl-*cis,cis*-cycloocta-1,5-dienes by MIR with 1-methyl-*cis,cis*-cycloocta--1,5-diene and 3-methyl-*cis,cis*-cycloocta-1,5-diene. As seen in Figure 8, only four common isomers are formed by the two methyl-cyclooctadienes. In addition in Figure 8 the catalytic preparation of the individual isomers by means of cyclodimerisztion of isoprene or piperylene or cross-dimerization of butadiene with dimethyl- or ethylsubstituted 1,3-dienes is shown. It was thus for example possible to identify the cross-cyclodimers of isoprene and piperylene, a normally difficult task without MIR. In Figure 8 are given only those catalytic reactions carried out by us. The following experimental procedure was adopted: the nickel ligand catalyst was dissolved in the substituted 1,3-diene and butadiene was added at normal pressure and 30-60 °C as it was consumed [6, 7]. After completion of the reactions, excess of 1,3-dienes was distilled off, with the result that any four-membered ring products were catalytically rearranged to the corresponding *cis,cis*-cycloocta-1,5-dienes (see chapter 6.1.). The yields are generally around 10 to 40% based on converted butadiene and ca. 80-90% based on converted substituted 1,3-dienes.

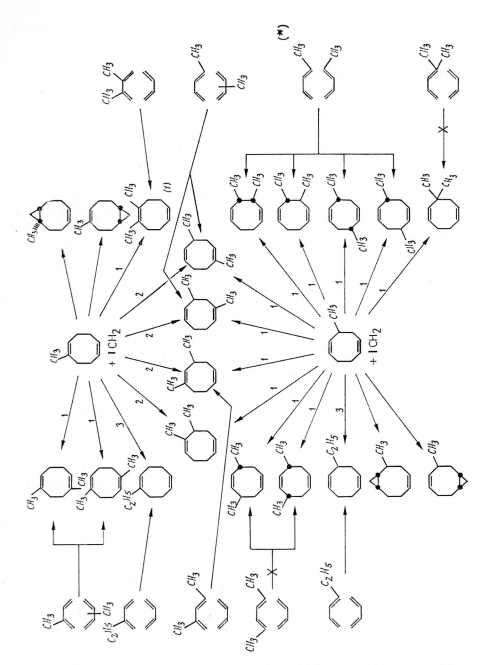

Figure 8. MIR-products of 1- and 3-methyl-*cis,cis*-cycloocta-1,5-diene. In addition the possible catalytic synthesis is shown.

(*) See chapter 6.1.

4.2. Preparation and identification of certain dimethyl *cis,trans*-cyclodeca-1,5-dienes

In chapter 2.3. we reported that the ten-membered ring Xb, which is formed together with two other isomers Xa and Xd from butadiene, isoprene and ethylene using a nickel-ligand catalyst, can be isolated in a pure form. By a lucky chance, the cyclodeca-1,5-diene Xa was identified as the only cyclic cross-cotrimer from butadiene, isoprene and ethylene using a nickel catalyst without a phosphane or phosphite. Xa was purified by a thorough fractionation at 0,1 torr. Thus one could easily identify [24] the ten-membered rings prepared catalytically from two molecules of isoprene and one molecule of ethylene as derivatives of *cis,trans*-cyclodeca-1,5-diene. MIR was carried out with Xa and Xb. In Figure 9 are shown only those methylation products which can be detected in the MIR mixture with the statistical weight of one. As well as the isomers XIIa, XIIb and XIIc, XIId, which contain one methyl group per double bond, a fifth isomer XXIV is formed from both the monomethyl-cyclodecadienes. The 1,2-dimethyl-*cis,trans*-cyclodeca-1,5-diene is formally one of the two possible cyclic isomers, which could be formed in a catalytic cross-co-cyclotrimerization

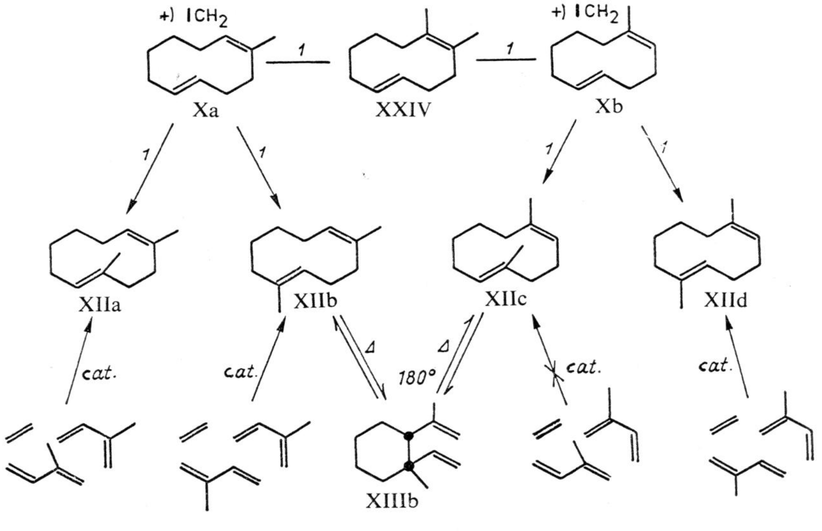

Figure 9. Syntheses of dimethyl-*cis,trans*-cyclodeca-1,5-dienes, substituted only on the double bonds, via nickel catalyzed reactions, MIR's of monomethyl-substituted cyclodecadienes and retro Cope Rearrangement of a corresponding six-membered ring [24].

of 2,3-dimethyl-butadiene, butadiene and ethylene. As stated in chapter 2.3. and 6.2. only XIIa, XIIb and XIId of the four possible isomeric ten-membered rings are formed catalytically from isoprene and ethylene. The fourth isomer XIIc was synthesized by two independent routes. Firstly by heating the six-membered ring XIIIb to 180 °C and secondly by means of the MIR just described

Table 3

KOVATS INDICES OF DIMETHYL-CIS,TRANS-CYCLODECA-1,5-DIENES SUBSTITUTED ONLY ON THE DOUBLE BONDS [24]

Compound	Kovats indices average value	MIR				Catalytically formed dimethylcyclodecadienes			
		Xa	Δ	Xb	Δ	$2\, \diagup\!\!\diagdown\!\!\diagup + C{=}C$	Δ	XIIIb	Δ
XIIa	1256.1	1256.3	+ 0.2			1255.9	− 0.2		
XIIb	1262.6	1262.8	+ 0.2			1262.7	+ 0.1	1262.3	− 0.3
XIIc	1264.0			1264.1	+ 0.1			1263.9	− 0.1
XIId	1268.2			1268.3	+ 0.1	1268.0	− 0.2		
XXIV	1272.5	1272.4	− 0.1	1272.6	+ 0.1				

(see Figure 9). Sections of all relevant gas-chromatograms are depicted in Figure 10. Only the peaks of those compounds which are formed with the sta-

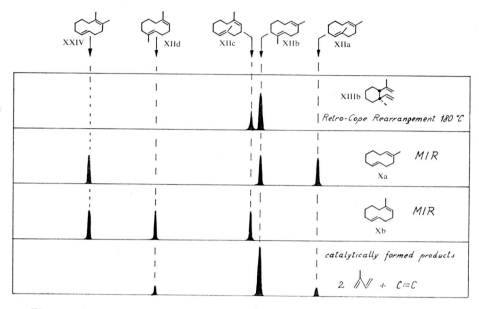

Figure 10. Sections of the gas-chromatograms of dimethyl-*cis,trans*-cyclodeca-1,5-dienes [24].

tistical weight of one in the MIR mixtures are reproduced in the middle section of Figure 10. The complementary Kovats indices of these individual isomers are summarized in Table 3.

4.3. Preparation and identification of certain dimethyl derivatives of ttt- and ttc-cyclododeca-1,5,9-triene

The identification of the dimethyl-cyclododeca-1,5,9-trienes, which are formed on chromium, titanium and nickel containing catalysts from one molecule of butadiene and two molecules of isoprene or from butadiene and 2,3-dimethyl-butadiene in a 2 : 1-ratio respectively (see chapter 6.3.), requires the isolation of all 1-methyl-cyclododeca-1,5,9-trienes with various configurations. The all-*cis*-isomer of the unsubstituted and the 1-methylsubstituted twelve-membered ring are not formed catalytically. The corresponding isomers with a tcc-configuration arise in low yields only when using a low valent nickel catalyst (see chapter 6.3.). Therefore we investigated those dimethylsubstituted isomers with a ttt- and a ttc-configuration in the twelve-membered ring skeleton [41, 42]. The following Kovats indices of the 1-methyl-cyclododeca-1,5,9-trienes (1-Me-CDT) were determined using a 100 m capillary column with squalane as liquid phase at 120 °C [32]:

1-M-ttt-CDT	(XXVa)	$I^S_{120} = 1345$
1-M-ctt-CDT	(XXVb)	$I^S_{120} = 1353$
1-M-tct-CDT	(XXVc)	$I^S_{120} = 1364$
1-M-ttc-CDT	(XXVd)	$I^S_{120} = 1364$

The ttt-, ctt-isomer and the tct- and ttc-mixture can be separated by preparative scale gas-chromatography in greater than 99% purity and in 50-70% yield. The gas-chromatograms of the MIR mixtures of all isomers are shown in Figure 12. The corresponding reaction schemes are depicted in Figure 11 and 13.

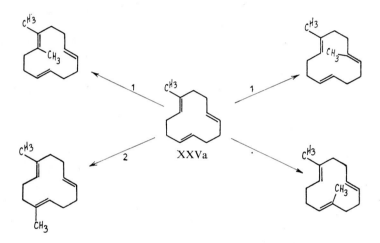

Figure 11. All dimethyl-ttt-cyclododeca-1,5,9-trienes with both methyl groups on the double bonds prepared by MIR [41].

We learned from these gas-chromatographic studies that using a titanium catalyst only the 1-Me-ttc-CDT is formed, whereas using a chromium catalyst a mixture of 95% 1-Me-tct- and 5% 1-Me-ttc-CDT is obtained. The identification of these two isomers was only possible once the positions of the methyl groups in the dimethyl-derivatives shown in Figure 13 were known. The identification procedure will be exemplified in the cases of the isomers XXVIa, XXVIb and XXVIc. These are formed from only one of the three 1-methyl-cyclododeca-1,5,9-trienes (XXVb, XXVc or XXVd). The identification requires information from the catalytic reactions. XXVIc is synthesized in a catalytic cross-cyclotrimerization of butadiene and 2,3-dimethyl-butadiene (2 : 1). XXVIa could be enriched to 80% using preparative gas-chromatography. After hydrogenation of XXVIa we could identify the products formed as *cis*- and *trans*-iso-

106

mers of 1,6-dimethylcyclododecanes (see next chapter). This hydrogenation can be carried out excellently (after separating the individual olefinic isomers in a capillary column) directly before their mass spectra are measured. Using in these cases the hydrocarbon mass spectra as fingerprints, which must be known from

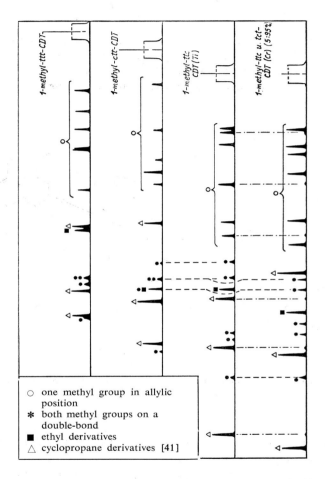

Figure 12. MIR products of 1-methyl-cyclododeca-1,5,9-trienes liquid phase: squalane 120 °C.

independent experiments, their structures can normally be identified. An application of this combination of capillary gas-chromatography with mass spectrometry and hydrogenation after the separation [41] is described for one example in chapter 5. The exact identification of all dimethyl-cyclododecatrienes was only possible by determining the retention data of the individual isomers in two

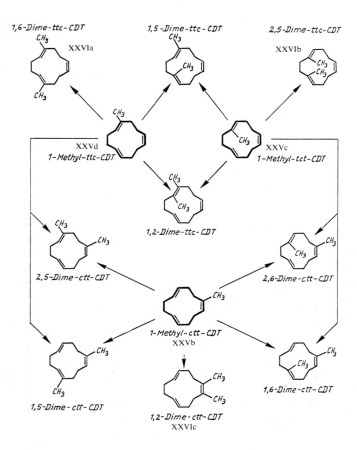

Figure 13. All dimethyl-cyclododeca-1,5,9-trienes with a ttc-configuration and with both methyl groups on the double bonds, prepared by MIR from 1-Me-ctt-, 1-Me-ttc- and 1-Me-tct-CDT [41].

columns with different liquid phases, methylsilicon-oil DC 200 and squalane at 120 °C being used.

From the MIR's of all 1-methyl-cyclododeca-1,5,9-trienes we also obtained the retention data of the corresponding 1-ethyl-cyclododeca-1,5,9-trienes, which are formed with the statistical weight of three.

4.4. Preparation and identification of all dimethylcyclododecanes

In Table 4 are listed the Kovats indices (DC 200 at 120 °C) of all dimethyl--cyclododecanes, which are prepared by MIR of methylcyclododecane and hydrogenation of catalytically formed cross-cyclotrimers of butadiene and isoprene

(1 : 2) or butadiene and 2,3-dimethyl-butadiene (2 : 1) [42, 43]. In addition, we have hydrogenated the cross-cyclotrimers of butadiene and piperylene (1 : 2).

Table 4

THE KOVATS INDICES (CAPILLARY COLUMNS DC200, 120 °C) OF DIMETHYLCYCLODODE-CANE ISOMERS FORMED BY MIR OF METHYLCYCLODODECANE AND HYDROGENATION OF CROSS-CYCLOTRIMERS FROM BUTADIENE AND TWO ISOPRENE MOLECULES OR 2,3-DIMETHYL--BUTADIENE AND TWO BUTADIENE MOLECULES, RESPECTIVELY [41]

Dimethyl-cyclododecane	Average-value Kovats index	methylcyclododecane		catalytically formed dimethyl-cyclododeca-1,5,9-triene isomers, (hydrogenated)					
		MIR	Δ	Ni	Δ	Cr	Δ	Ti	Δ
cis-1,3-		1422.6							
trans-1,4-	1425.4	1425.4	± 0.0	1425.2	— 0.2	1425.2	— 0.2	1425.7	+ 0.3
trans-1,3-		1428.4							
cis-1,5-	1430.6	1430.5	— 0.1	1430.4	— 0.2	1430.7	+ 0.1	1430.7	+ 0.1
trans-1,6-	1432.3	1432.3	± 0.0	1432.1	— 0.2	1432.3	± 0.0	1432.5	+ 0.2
trans-1,5-	1437.5	1437.5	± 0.0	1437.3	— 0.2	1437.5	± 0.0	1437.6	+ 0.1
cis-1,5-	1438.8	1438.9	+ 0.1	1438.7	— 0.1	1438.7	— 0.1	1439.0	+ 0.2
cis-1,7-		1439.7							
cis-1,4-	1447.6	1447.6	± 0.0	1447.5	— 0.1	1447.6	± 0.0	1447.8	+ 0.2
trans-1,7-		1449.4							
trans-1,2-	1457.3	1457.4	+ 0.1	1457.2	— 0.1				
cis-1,2-	1459.0	1459.0	± 0.0	1459.0	± 0.0				

The *cis,trans*-isomer-pairs of 1,2-, 1,3-, 1,4-, 1,5- and 1,6-dimethylcyclododecane arise with the statistical weight of two, whereas the two 1,7-isomers and the 1,1-isomer can be detected with the statistical weight of one. Catalytic hydrogenation of the cross-cyclotrimers of butadiene and 2,3-dimethylbutadiene (2 : 1) yields only the 1,2-pair. The hydrogenation-products of the cross-cyclotrimers from butadiene and piperylene (1 : 2) consist of a mixture of 1,2-, 1,5 and 1,6-dimethylcyclododecanes, whereas a mixture of 1,4-, 1,5- and 1,6-isomers is obtained starting from butadiene and isoprene. The 1,5- and 1,6-dimethylcyclododecanes could be differentiated by hydrogenation of enriched 1,6-dimethyl-ttc-cyclododeca-1,5,9-triene (XXVIa), as described in the foregoing chapter. Knowing all retention data of the dimethylcyclododecanes prepared, it is easy to identify for example after hydrogenation the structures of some byproducts in the catalytic co-oligomerization of butadiene with allene (see chapter 6.2.). The assumed configurations of the individual dimethylcyclododecanes listed in Table 4 is reasonable but not based on experimental evidence.

5. IDENTIFICATION OF ISOMERS USING THE COMBINATION GAS CHROMATOGRAPHY-MASS SPECTROMETRY

In the presence of morpholine, butadiene is almost exclusively dimerized on a nickel triethylphosphite catalyst to *cis,trans*- and *trans,trans*-octa-1,3,6-triene (XXVII) and (XXVIII) [44, 45]. When conversion is greater than 80% octatrienes formed codimerize with butadiene to form *n*- and *iso*-C_{12}-tetraenes. The *iso*-C_{12}-tetraenes XXIXa, XXIXb and XXIXc are nearly statistically formed [45] (*) (for the mechanism see chapter 7. equations 42 and 46).

$$(19)$$

In addition, the cross-dimerization of butadiene and isoprene was investigated [45]. All eight possible isomers could be identified by their retention data form MIR mixtures of *cis,trans*- and *trans-trans*-octa-1,3,6-triene. The pertinent sections of the chromatograms of the MIR mixtures and the seven monomethyl--octa-1,3,6-trienes, prepared in the catalytic cross-dimerization, are shown in Figure 15. The *cis*- or *trans*-configuration of the methyl-octa-1,3,6-trienes in the 3-position of the chain is evident from the retention index of the individual isomer (see Figure 14). By gas chromatography-mass spectrometry and subsequent hydrogenation of the isomers after separation, an unambiguous analysis of the carbon skeleton was possible. The detailed structures of the isomers further enriched by preparative gas-chromatography could be assigned using ^{1}H--NMR-spectroscopy (see also chapter 7. equations 42 and 46).

Reaction of 270g butadiene and 340g isoprene (molar ratio 1 : 1) with 17 mmole of the catalyst (Ni : P(C_6H_5)$_3$: morpholine $= 1 : 1 : 70$) for 2,5 hours yields the monomethyl-octa-1,3,6-trienes as 25% of the reaction-products (conversion: $\sim 40\%$ butadiene and $\sim 10\%$ isoprene). The percentage composition of the methyl-octa-1,3,6-triene ($\Sigma = 100\%$) is given in Figure 15.

(*) The unsaturated alkylamines, which were isolated at lower temperatures in this reaction [44], are not formed from butadiene and morpholine on low valent nickel ligand catalysts as it is assumed in this paper. This was shown by W. Fleck [45]. Whether these amines are formed by aluminium or nickel catalysts of a higher oxidation-state, which were perhaps present as impurities, is under investigation.

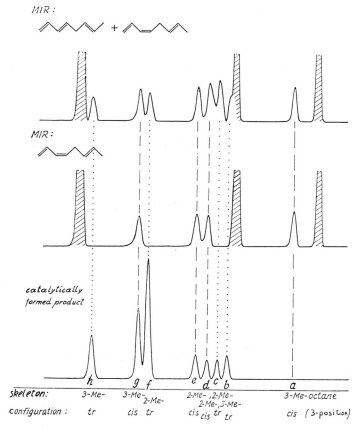

Figure 14. **MIR** products of *trans,trans-* and *cis,trans-*octa-1,3,6-trienes and open chain cross-dimers catalytically formed, from butadiene and isoprene [45].

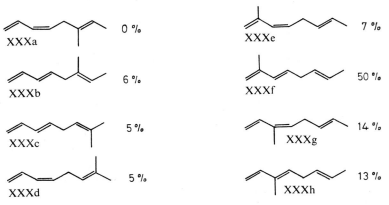

Figure 15. Percentage distribution of the methyl-octa-1,3,6-trienes catalytically formed from butadiene and isoprene [45].

6. CATALYTIC SYNTHESES OF MONO AND DIMETHYL SUBSTITUTED RING OLEFINS

From investigations of the catalytic cyclodimerization of butadiene it is known that, beside 60% of *cis,cis*-cycloocta-1,5-diene, 40% of *cis*-1,2-divinylcyclobutane are formed as kinetically controlled product. Greater than 85% conversion of butadiene causes a decrease in *cis*-1,2-divinylcyclobutane. Further, after total conversion of the 1,3-diene, no four-membered ring can be detected. This is due to the following catalytic "Cope Rearrangement" [20]:

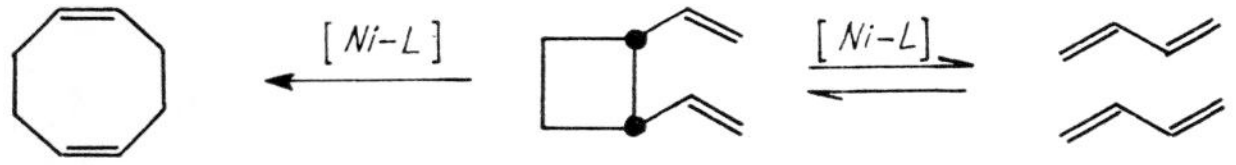

$$(20)$$

A Cope-Rearrangement reaction catalyzed by a transition metal complex has been described by Trebellas, Olechowsky and Jonassen [46]:

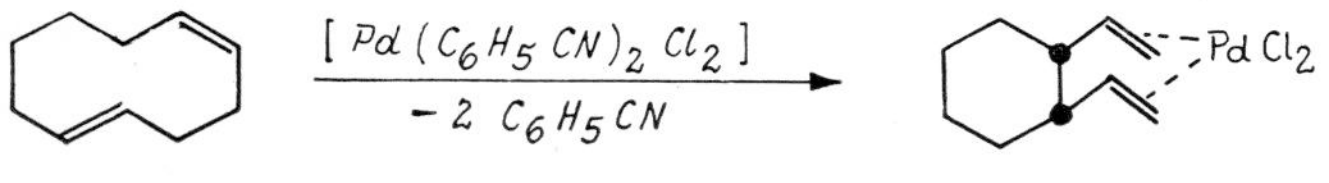

$$(21)$$

In this case the rearrangement is stoichiometric with respect to the complex. The above "catalytic Cope Rearrangement" of *cis*-1,2-divinylcyclobutane, which is catalytic with respect to the nickel can easily be recognized from the volume--contraction against time curves of the catalytic and thermal Cope Rearrangement of the four- to the eight-membered ring obtained by dilatometry (see Figure 16, [18]):

cis-1,2-divinylcyclobutane	$d_4^{20} = 0,80$
cis,cis-cycloocta-1,5-diene	$d_4^{20} = 0,88$
butadiene	$d_4^{20} = 0,62$

The thermal reaction is first order in *cis*-1,2-divinylcyclobutane whereas the catalytic rearrangement is zero order with respect to the cyclobutane derivative. The latter, however, is dependent on the concentration of the nickel ligand catalyst and on the nature of the substituents in the phosphorus-containing ligand. Surprisingly, in the first few minutes of the "catalytic Cope Rearrangement" a volume expansion instead of contraction is observed. This is due to the fact that the four-membered ring is in an equilibrium, promoted by the catalyst, with 3-4% of butadiene. As seen in Figure 16, the equilibrium is temperature dependent. When the butadiene is removed from the reaction mixture under reduced pressure, more than 30% butadiene, based on converted *cis*-1,2--divinylcyclobutane, can be isolated.

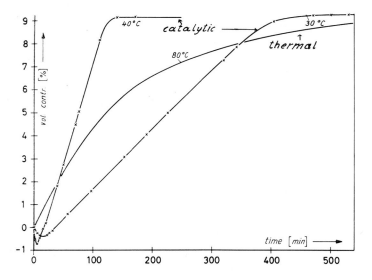

Figure 16. Volume contraction against time curves of thermal and « catalytic » Cope-Rearrangement of *cis*-1,2-divinylcyclobutane to *cis,cis*-cycloocta-1,5-diene (Ni : L : *cis*-1,2-divinylcyclobutane = = 1 : 1 : 50 : L = tri(*o*-phenylphenyl)phosphite) [18].

6.1. Catalytic cyclodimerization of piperylene

By using pure *cis*-, pure *trans*- and *cis,trans*-mixtures of piperylene we hoped to gain informations concerning the stereochemical course of the carbon-carbon bond formation in the catalytic cyclodimerization of 1,3-dienes. In all experiments nickel-ligand catalysts with a 1 : 1 Ni-ligand ratio were used. Nickel-bis-cycloocta-1,5-diene was employed as the low-valent nickel complex. The catalyst: 1,3-diene ratio was 1 : 50. 20% *n*-octane was added as an internal standard. After deactivating the catalyst (by addition of three mmoles triphenylphosphite per mmole catalyst and distillation of the volatile reaction mixture followed by gas chromatographic analysis) the concentration against time curves of all products could be obtained [18, 21, 47].

Assuming that only dimethyl-derivatives with *cis*-1,2-divinylcyclobutane- and *cis,cis*-cycloocta-1,5-diene skeletons are formed, then ten dimethyl-1,2-divinylcyclobutanes and four dimethyl-*cis,cis*-cycloocta-1,5-dienes should result. The following four-membered rings could not be detected in the reaction-mixtures:

| VIIf | VIIg | VIIh | VIIi | VIIj |

The piperylenes from which the individual four- and eight-membered rings originated are shown in Figure 17. The following catalysts were used in the experiments:

catalyst A : L = tri(*o*-phenylphenyl)phosphite
catalyst B : L = triphenylphosphane
catalyst C : L = tricyclohexylphosphane

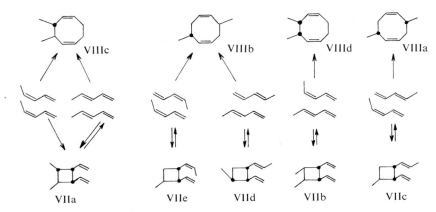

Figure 17. Structures and configurations of four- and eight-membered rings obtained from piperylene with various configurations [18].

In Figures 18 and 19 are depicted the concentration against time curves of the four- and the eight-membered rings obtained from *trans*-piperylene using

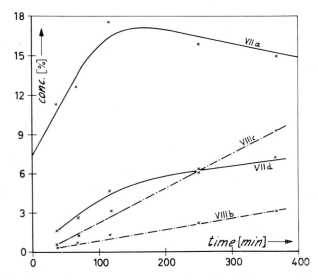

Figure 18. Concentration against time curves of the products formed from *trans*-piperylene by catalyst A at 30 °C [18].

catalysts A, B and C. The relative rates are influenced by the nature of the ligand. The formation of the eight-membered rings involves reactions which are zero order in *trans*-piperylene, as can be seen from Figure 18, where the concen-

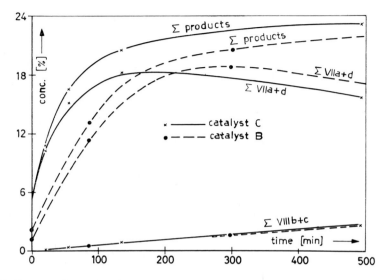

Figure 19. Concentration against time curves of the products formed from *trans*-piperylene by catalyst B and C at 30 °C [18].

tration against time curves of the individual isomers are drawn. The rate of *cis*-1,2-divinylcyclobutane formation from butadiene is zero order up to 75% 1,3-diene conversion. The concentration against time curves of the dimethyl-*cis*-1,2-divinylcyclobutane VIIa, however, reach a maximum at relatively low conversion of the piperylene.

6.1.1. TEMPERATURE DEPENDENT EQUILIBRIUM BETWEEN TRANS-PIPERYLENE AND THE CORRESPONDING CIS-1,2-DIVINYLCYCLOBUTANE

The concentration maximum of VIIa can be unequivocally interpreted by assuming an equilibrium between *trans*-piperylene and the four-membered ring VIIa [18].

$$K = \frac{[\text{four-membered ring VIIa}]}{[trans\text{-piperylene}]^2}$$

This equilibrium does not lie so much in favour of the four-membered ring as in the case of butadiene (see equation 20 and Figure 15 in chapter 6.). As soon as VIIa reaches its concentration maximum, K has (within experimental error) a value independent of the nature of the catalyst (K = 2.3 ± 0.3 × 10^{-2} (mole/l)^{-1}).

The temperature-dependence of the equilibrium can easily be demonstrated by the temperature-jump method (see Figure 20).

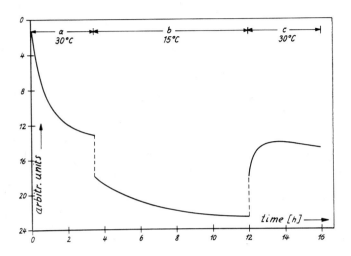

Figure 20. Volume expansion and contraction of the reaction mixture from *trans*-piperylene and a nickel tricyclohexylphosphane catalyst at successive temperatures of 30, 15 and 30 °C [18].

trans-piperylene was reacted using catalyst C, because, apart from the four-membered ring VIIa, all other products are formed extremely slowly (see Figure 19). The attainment of the equilibrium at 30 °C (Figure 20, section a: contraction) followed by temperature-jump to 15 °C (section b: further contraction) and temperature-jump back to 30 °C (section c: expansion) is shown in Figure 20. The exact original volume is not reached, because other products are formed.

6.1.2. CATALYTIC ISOMERIZATION OF *cis*- TO *trans*-PIPERYLENE

The interpretation of the concentration against time curves of the catalytic cyclodimerization of pure *cis*- and mixtures of *cis*- and *trans*-piperylene is complicated by the fact that *cis*-piperylene is catalytically isomerized to *trans*-piperylene. This isomerization occurs more rapidly with catalyst C than with catalyst B or A [18]. In the reaction mixtures there is an equilibrium between ~ 64% *trans*- and ~ 36% *cis*-piperylene. Because this equilibrium is quickly attained using catalyst C and the formation of the eight-membered rings is slow, it is understandable that, starting with pure *cis*-piperylene or a mixture of *cis*- and *trans*-piperylene corresponding to the above mentioned equilibrium, the percen-

116

tage composition of the various eight-membered rings should be nearly identical when the 1,3-dienes are totally converted (see Table 5). It is assumed [18, 47] that *cis*-piperylene isomerizes to the *trans*-1,3-diene via the four-membered ring VIIa and also that *trans*-piperylene isomerizes to *cis*-piperylene albeit to a lesser extent (see also Table 5).

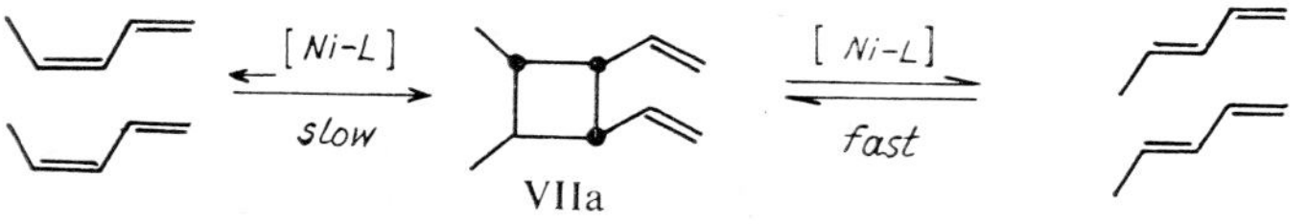

$$\text{(22)}$$

Table 5

PERCENTAGE COMPOSITION OF THE INDIVIDUAL EIGHT-MEMBERED RINGS ($\Sigma = 100\%$) AT 100% CONVERSION OF THE 1,3-DIENE DEPENDENT ON THE CONFIGURATION OF THE PIPE-RYLENE REACTED AND THE CATALYST USED. IN PARENTHESES ARE GIVEN THE YIELDS OF THE EIGHT-MEMBERED RINGS, BASED ON TOTALLY CONVERTED 1,3-DIENES [18]

Original piperylene	Eight-memb. ring	Catalyst A	Catalyst B	Catalyst C
cis-piperylene	VIIId	27%	8%	6%
	VIIIc	15% (90%)	36% (78%)	67% (77%)
	VIIIb	40%	16%	6%
	VIIIa	18%	40%	21%
trans-piperylene (+1.3% *cis*-piperylene)	VIIId	2%	2%	5%
	VIIIc	64% (96%)	64% (85%)	75% (75%)
	VIIIb	33%	27%	6%
	VIIIa	1%	8%	14%
cis- (34%) and *trans*- (64%) piperylene	VIIId	32%	8%	6%
	VIIIc	20% (91%)	37% (77%)	66% (70%)
	VIIIb	26%	16%	6%
	VIIIa	22%	39%	22%

The kinetic evidence will be discussed in detail in [48]. A further possible mechanism of the isomerization will be discussed in chapter 8.2.2. equation 59 and 60.

6.1.3. CATALYTIC COPE REARRANGEMENT AND CLEAVAGE OF THE DIMETHYL-1,2-DIVINYLCYCLOBUTANES TO 1,3-DIENES

As mentioned in chapter 6. *cis*-1,2-divinylcyclobutane is rearranged to *cis*, *cis*-cycloocta-1,5-diene by nickel-ligand catalysts. At the same time there is an equilibrium between the diene and the dimethyl-divinylcyclobutanes; both of these reactions have also to be considered [18].

The catalytic cleavage of the individual dimethyl-*cis*-1,2-divinylcyclobutanes to piperylenes having *cis*- or *trans*-configuration is the simplest way to show (by using the principle of microscopic reversibility) from which piperylenes the four-membered rings are formed (see Figure 17). Normally this procedure, as in the case of the catalytic syntheses of the dimethyl-*cis*-1,2-divinylcyclobutanes, is complicated by the isomerization of *cis*-piperylene and the equilibria between 1,3-dienes and various cyclobutane derivatives. A clear cut decision could be made when carrying out the cleavage of the four-membered rings under reduced pressure, so as to immediately remove the 1,3-diene from the system. In addition, the best results were obtaines using tri(*o*-phenylphenyl)-phosphite as the ligand for nickel. In Table 6 is shown the piperylene isomer distribution when VIIe is split into the 1,3-dienes and rearranged to VIIIb using various catalysts under various conditions. Using catalyst A at 30 to 50 torr and rearranging all isolated *cis*-1,2-divinylcyclobutane derivatives, we found the stereochemical results given in Figure 17.

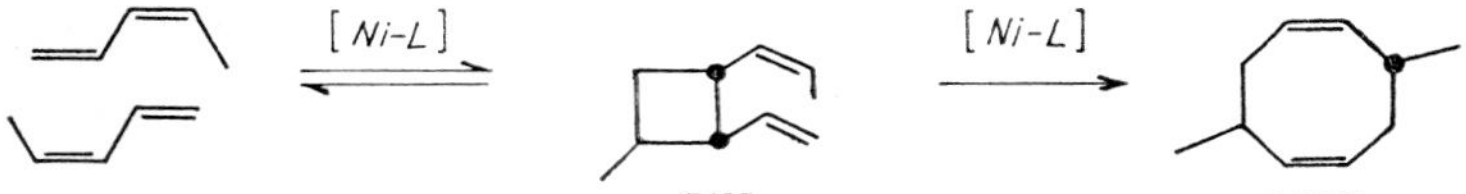

$$(23)$$

$$\text{VIIe} \qquad \text{VIIIb}$$

Table 6

PERCENTAGE DISTRIBUTION OF *cis*- AND *trans*-PIPERYLENE FROM THE CLEAVAGE OF THE FOUR-MEMBERED RING VIIe, DEPENDENT ON THE CATALYST AND THE REACTION CONDITIONS [18]

Reaction conditions	Isolated piperylene isomers	Catalyst used		
		A	B	C
Normal pressure	*cis*-	88 %	70%	33%
	trans-	12 %	30%	67%
Vacuum 30-50 torr	*cis*-	99,4%	95%	90%
	trans-	0,6%	5%	10%

6.2. Catalytic syntheses of *cis,trans*-cyclodeca-1,5-diene derivatives

This chapter will be mainly concerned with the catalytic synthesis of mono- and dimethyl substituted *cis,trans*-cyclodeca-1,5-dienes. In addition two related reactions of this type will be mentioned. To synthesize the dimethyl derivatives the following mixtures were reacted with low-valent nickel catalysts:

1. isoprene and ethylene

2. *cis*-piperylene and ethylene

3. *trans*-piperylene and ethylene

4. mixture of *cis*- and *trans*-piperylene and ethylene

5. butadiene and *trans*-but-2-ene

6. butadiene and iso-butene.

The monomethyl-derivatives were obtained by cross-co-cyclotrimerization and co-cyclotrimerization from:

1. butadiene and propylene

2. butadiene, isoprene and ethylene

3. butadiene, piperylene (*cis*-, *trans*- and *cis*,*trans*-mixtures) and ethylene.

The independent syntheses by MIR's and the identification by Cope Rearrangements of most of these *cis*, *trans*-cyclodeca-1,5-diene derivatives are described in chapters 2.3., 3.2. and 4.2..

6.2.1. CATALYTIC CO-TRIMERIZATION OF ISOPRENE AND ETHYLENE

In all experiments standard conditions were maintained, as in the case of the co-cyclotrimerization of butadiene and ethylene [11]. The reaction temperature was normally 40 °C. The initial molar ratio of ethylene to isoprene was 1 : 1 and the nickel to ligand ratio was also 1 : 1. The reactions were carried out in autoclaves.

The co-trimerization of isoprene and ethylene was the most extensively investigated reaction [24, 39]. Table 7 shows the experimental data obtained, using a variety of nickel catalysts both with and without phosphorus containing ligands. The following products are formed:

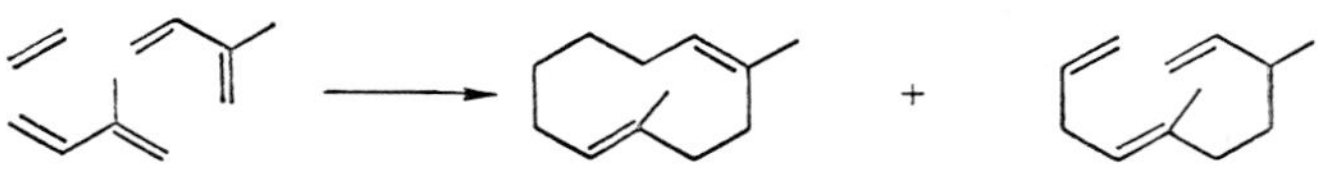

XIIa + XXXIb (24)

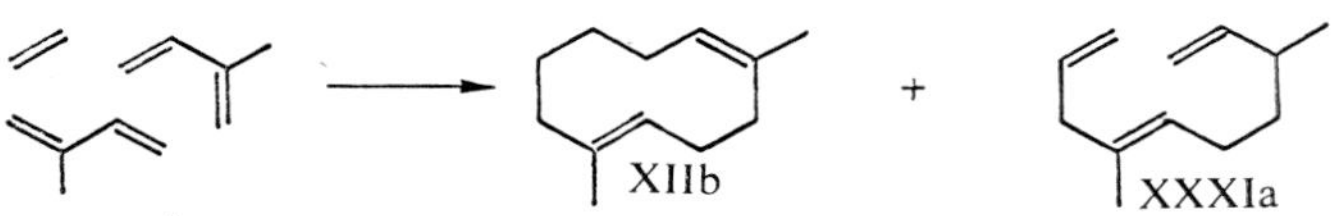

XIIb + XXXIa (25)

XIIc + XXXId (26)

Table 7

THE DEPENDENCE OF THE PRODUCT DISTRIBUTION ON THE LIGAND IN THE CO-TRIMERIZATION OF ISOPRENE AND ETHYLENE; Ni : LIGAND = = 1 : 1; 40 °C; INITIAL ISOPRENE : ETHYLENE RATIO = 1 : 1 [24]

Ligand	Without	$P(C_6H_{11})_3$	$P(OC_2H_5)_3$	$P(C_6H_5)_3$	$P(OC_6H_5)_3$	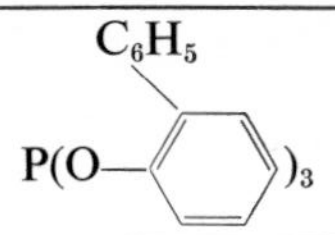 $P(O\text{-}C_6H_4\text{-}C_6H_5)_3$
Cyclic dimers of isoprene	2.8%	2.8%	2.5%	2.9%	3.4%	15.4%
Σ?	4.1%	0.6%	0.6%	0.1%	0.2%	0.1%
XXXIb	19.6%	40.7%	23.4%	13.8%	8.4%	0.8%
XXXIa	35.5%	8.4%	17.1%	12.6%	9.8%	0.3%
XXXId	13.0%	17.2%	24.0%	18.6%	25.9%	24.4%
XXXIc	4.5%	6.3%	4.6%	8.4%	5.5%	3.5%
XIIa	2.3%	0.4%	1.1%	1.4%	2.7%	1.9%
XIIb	9.1%	16.9%	19.9%	33.8%	40.7%	31.4%
XIId	0.1%	—	0.2%	0.1%	0.8%	8.4%
Higher oligomers	9.0%	6.7%	6.6%	7.3%	2.6%	3.8%
g product/g Ni · day	27.1	22.8	37.6	40.0	42.4	21.2
Conversion isoprene	79 %	65 %	71 %	56 %	41 %	56 %
Conversion ethylene	32 %	28 %	31 %	24 %	18 %	20 %

$$\text{(27)}$$

Only the cyclic isomer XIIc is not catalytically synthesized. The following characteristic differences in the catalytic reaction of isoprene and ethylene and the co-trimerization of butadiene and ethylene were found:

1. The open-chain and cyclic co-trimers of isoprene and ethylene are formed in 80 to 90% yield in the reaction mixture. Only 2-3% yields of cyclic isoprene dimers could be detected (exception, L = tri(o-phenylphenyl) phosphite : : 15%). Under similar reaction conditions, using a butadiene to ethylene ratio of 1 : 1, 25% (L = tricyclohexylphosphane) to 94% (L = tri(o-phenylphenyl) phosphite) of cyclic butadiene dimers are still produced.

2. The proportion of the open-chain isomers XXXIa-d is from 48% (L = tri(o-phenylphenyl)phosphite) to 78% (L = tricyclohexylphosphane) of all dimethyl-C_{10}-compounds formed. The corresponding percentages in the co-trimerization of butadiene and ethylene are less than 0.1% and 22%, respectively.

3. The rate of the catalytic co-trimerization of 1,3-dienes and ethylene is much smaller in the case of isoprene than butadiene. This difference in rate depends on the additional ligand on nickel.

In all the experiments summarized in Table 7, the conversion of isoprene lies between 41 and 71%. Using a nickel-triphneylphosphite catalyst the percentage composition of the products as a function of the isoprene conversion (11-86%) was investigated. In this case it was found that whether 40 or 70% of the 1,3-diene is converted, the percentage composition is hardly influenced. Therefore, it is reasonable to assume that the differences in the yields of the individual products are due to the additional ligand on the nickel and are not related to the amount of isoprene converted. The dimethyl-cis,trans-cyclodeca-1,5-dienes XIIa, XIIb and XIId were identified as described in chapter 2.3. and 4.2. By exhaustive fractionation the open-chain isomers XXXIa, XXXIb and XXXId could be isolated in more than 75% purity. Their structures could be proved by their IR- and ^{1}H-NMR spectra. All four isomers were hydrogenated and identified by comparing the retention data (obtained in capillary columns with polar and non polar phases) with the data of authentic samples of 2,6-, 2,7-, 3,6- and 3,7-dimethyl--decane.

6.2.2. CATALYTIC CO-OLIGOMERIZATION OF PIPERYLENE AND ETHYLENE

The co-trimerization of piperylene and ethylene was mainly investigated using a nickel-triphenylphosphite catalyst [24]. Using the same reaction conditions as in the case of isoprene and ethylene, pure cis-, nearly pure trans- and a *cis-trans*-mixture of piperylene were reacted with ethylene. The structures and configurations of the dimethyl-*cis,trans*-cyclodeca-1,5-dienes formed are shown in the next equations:

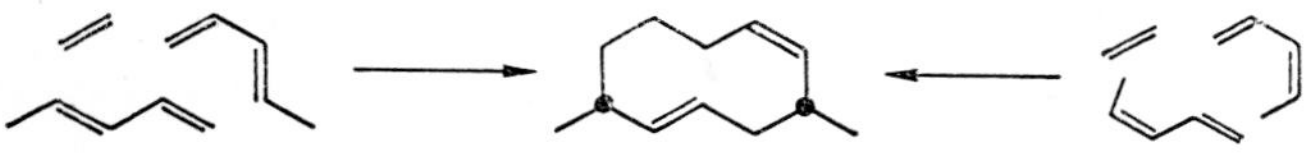

XVIIIa (28)

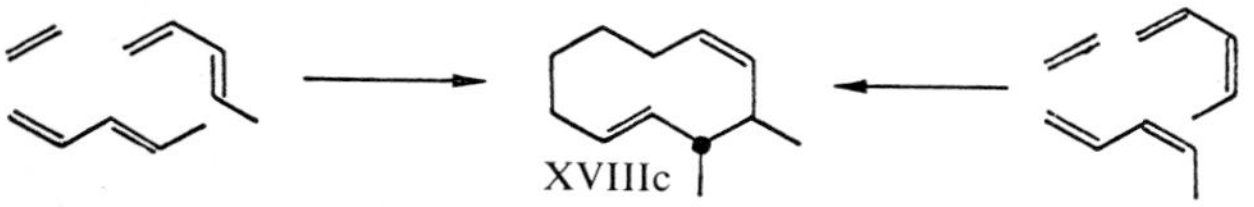

XVIIIc (29)

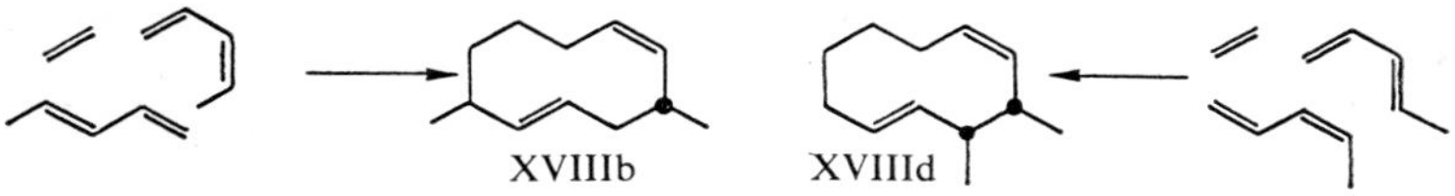

XVIIIb   XVIIId (30)

Table 8

PRODUCT DISTRIBUTION DEPENDENT ON THE CONFIGURATION OF THE PIPERYLENE; NI : L = = 1 : 1; REACTION TEMPERATURE 40 °C; INITIAL PIPERYLENE: ETHYLENE RATIO = 1 : 1 [24]

Composition of piperylene	98,7% *trans-* 1,3% *cis-*	100% *cis-*	62% *trans-* 38% *cis-*
Cyclic dimers of piperylene	19.3%	21.2%	16.4%
XVIIIa	67.9%	9.2%	39.6%
XVIIIb	2.2%	1.8%	17.2%
XVIIIc	5.9%	56.5%	8.3%
XVIIId	0.3%	1.5%	10.0%
Higher oligomers	2.3%	2.7%	3.3%
g product/g Ni · day	4.6%	2.1%	5.2%

Only traces of open-chain co-trimers are formed. The cyclic co-trimers XVIIIa-d are formed in 69 to 76% yield, based on converted piperylene. When pure *cis-* or pure *trans*-piperylene are reacted, the ten-membered rings XVIIIa and XVIIIc result almost exclusively. The XVIIIa : c-ratio is 92 : 8 using *trans-*, but 14 : 86 using *cis*-piperylene. From a *cis-trans*-mixture of piperylene the additional ten-membered rings XVIIIb and XVIIId are formed. The small proportions of XVIIIb and XVIIId obtained using *trans-* or *cis*-piperylene are perhaps due to impurities (1,3% *cis*-piperylene in the *trans*-isomer) or an isomerization of *cis*- to *trans*-piperylene (see chapter 6.1. equation 22 and chapter 8.2.2. equations 59 and 60).

The reaction of piperylene and ethylene at 40 °C is extremely slow when a nickel-triphenylphosphite catalyst is used. The various rates of formation of open-chain and cyclic substituted and unsubstituted co-trimers (ΣC_{10}) from various 1,3-dienes under similar conditions are:

	butadiene	isoprene	*trans*-piperylene	*cis*-piperylene
gΣC_{10}/g Ni · day	188.7	39.8	3.5	1.6

When a *trans*-piperylene ethylene mixture is reacted with a low-valent nickel catalyst without a phosphorus containing ligand, 13% of the product which is isolated is a twelve-membered ring:

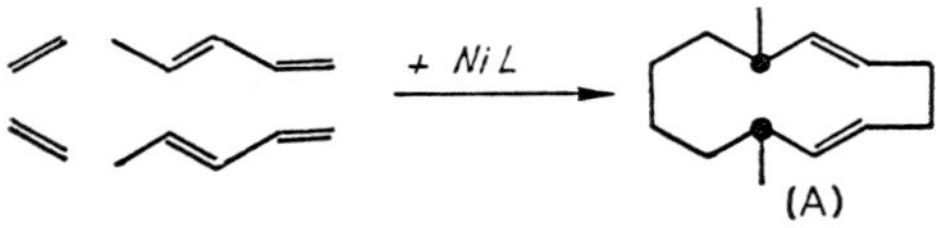

$$(31)$$

(A) was hydrogenated and gave one isomer of 1,6-dimethylcyclododecane, which is probably the *cis*-one.

To prepare the following dimethyl-*cis,trans*-cyclodeca-1,5-dienes, a 1 : 1--mixture of butadiene and isobutene or butadiene and *trans*-but-2-ene were heated at 40 °C in the presence of a low-valent nickel catalyst:

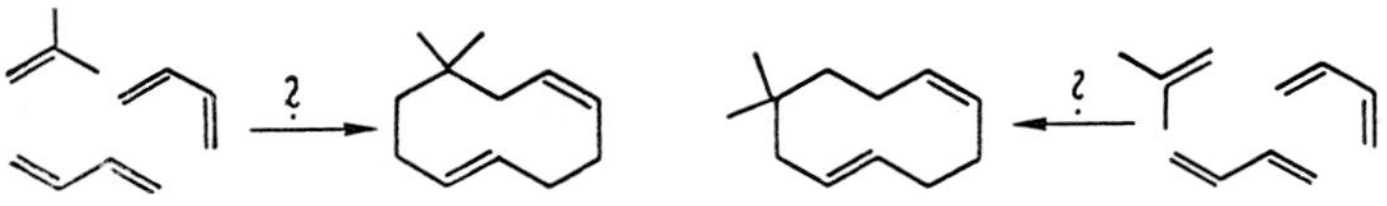

$$(32)$$

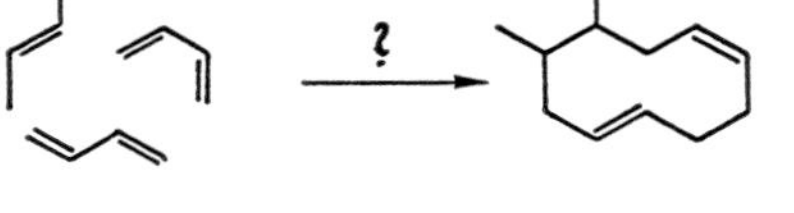

$$(33)$$

If ten-membered rings are formed, then it must be in less than 1% yield, as less than 1% of the products remained unidentified.

6.2.3. CATALYTIC CROSS-CO-CYCLOTRIMERIZATION OF BUTADIENE, ISOPRENE AND ETHYLENE

Table 9 shows the product distribution from the reaction of a 1 : 1 : 1 mixture of butadiene, isoprene and ethylene with a nickel-triphenylphosphite catalyst. Only the formation of mono- and dimethyl-substituted *cis,trans*-cyclodeca-

1,5-dienes will be discussed. The isomers of the ten-membered rings which are formed are shown in chapter 3.2. Figure 5. The experimental data are summarized in Table 9. Butadiene reacts much faster than isoprene. From the data in Table

Table 9

THE PRODUCT DISTRIBUTION AND PERCENTAGE CONVERSIONS OF THE 1,3-DIENES AND ETHYLENE IN THE CROSS-CO-OLIGOMERIZATION OF BUTADIENE, ISOPRENE AND ETHYLENE DEPENDENT ON THE REACTION TIME; REACTION TEMPERATURE 40 °C; L = TRIPHENYLPHOSPHITE; MOLAR RATIO BUTADIENE : ISOPRENE : ETHYLENE = 1 : 1 : 1 (AT THE BEGINNING OF THE REACTION) [24]

Reaction time	14h	24h	30h
Σ?	1.0%	1.7%	2.2%
Cyclic dimers of butadiene	56.4%	42.1%	39.4%
Cyclic crossdimers of but. + isoprene	15.1%	14.8%	15.0%
V	16.3%	16.3%	14.9%
ΣXa, Xb and Xd	11.0%	21.8%	22.7%
ΣXIIa, XIIb and XIId	0.3%	3.2%	4.6%
Conversion butadiene	61 %	79 %	83 %
Conversion isoprene	9 %	20 %	24 %
Conversion ethylene	8 %	18 %	19 %
g product/g Ni · h	4.0%	3.5%	3.0%

9 it is apparent that molar ratios of butadiene to isoprene in the range 1 : 2 to 1 : 4 favour the formation of cross-cyclodimers (mainly 1-methyl-*cis,cis*-cycloocta-1,5-diene) and monomethyl-*cis,trans*-cyclodeca-1,5 dienes (Xa, Xb, Xd, being formed in the ratio 39 : 39 : 22). Trace amounts of monomethyl-*trans*-deca-1,4,9-trienes occur and they could only be detected because their retention indices were known from MIR of *trans*-deca-1,4,9-triene (see chapter 3.2. Figure 6). As the isoprene concentration increases, so does that of the dimethyl-C_{10}-compounds with a subsequent decrease in the rate. When the reaction is catalyzed by a phosphane- or phosphite free low-valent nickel catalyst, more open-chain co- and cross-cotrimers are formed (ca. 7%), whereas only the cyclic cross-cotrimers Xa occurs as 20% of the total products.

6.2.4. CATALYTIC CROSS-CO-CYCLOTRIMERIZATION OF BUTADIENE, PIPERYLENE AND ETHYLENE

When 4% of *cis*- or *trans*-piperylene are added to a 1 : 1-mixture of butadiene and ethylene, the monomethyl-*cis,trans*-cyclodeca-1,5-diene XIVb, XIVc and XIVd form as 2-3% of the reaction mixture (see chapter 2.3. Figure 5).

Table 10

ISOMER DISTRIBUTION OF MONOMETHYL-*cis*, *trans*-CYCLODECA-1,5-DIENES CATALYTICALLY
FORMED FROM PIPERYLENE, BUTADIENE AND ETHYLENE

	cis-piperylene	*trans*-piperylene
XIVb	18%	36%
XIVc	44%	31%
XIVd	38%	33%

Under these reaction conditions 90% conversion of butadiene occurs and no dimethyl-*cis,trans*-cyclodeca-1,5-dienes are formed. Taking the sum of all cyclic C_{11} compounds as 100%, the percentage distribution of the individual isomers shown in Table 10 are obtained. Whether *cis*- or *trans*-piperylene is reacted hardly influences the relative proportions of the three isomers formed.

For preparative scale reaction the following procedure was adopted: the nickel-triphenylphosphite catalyst (85 mmole) was dissolved in 500 g of a 10 : 1 piperylene-butadiene mixture and transferred to an autoclave. After addition of 200 g of ethylene during 48 hours a further 800 g of butadiene were slowly injected at 40 °C. Of the ca. 1000 g of products 21% was present as cross-co-trimers of butadiene, piperylene and ethylene. Only three percent of these cross-cotrimers were identified as open-chain products (XXIIa, XXIIb and XXIIc).

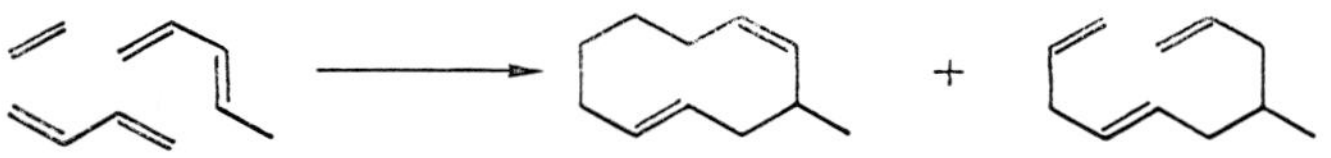

XIVb   +   XXIIa   (34)

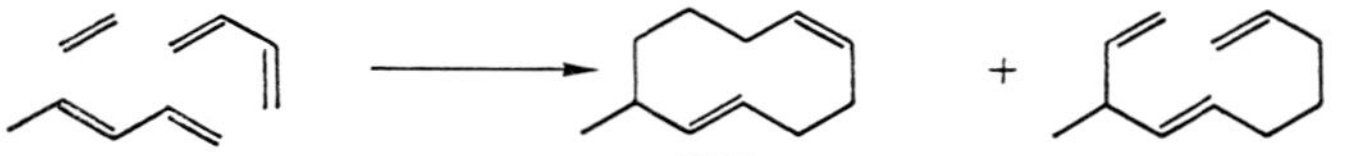

XIVc   +   XXIIb   (35)

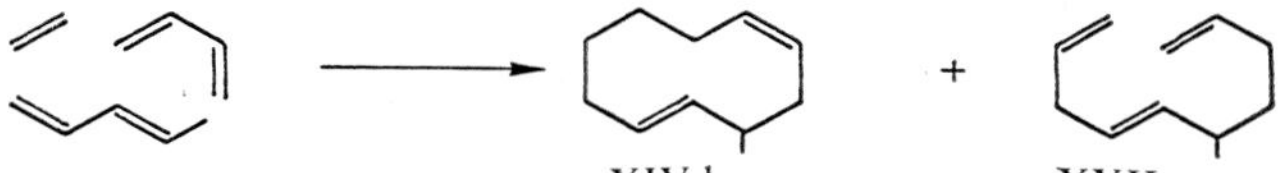

XIVd   +   XXIIc   (36)

6.2.5. CATALYTIC CO-TRIMERIZATION OF BUTADIENE AND PROPYLENE

The experimental data obtained in the co-trimerization of butadiene and propylene are reported in Table 11.

Table 11

THE PRODUCT DISTRIBUTION IN THE CATALYTIC CO-TRIMERIZATION OF BUTADIENE AND
PROPYLENE AT 40 °C [24]

| Butadiene : propylene | 1 : 1 | 1 : 2 | 1 : 1 |
ligand	$P(OC_6H_5)_3$	$P(OC_6H_5)_3$	without
Cyclic di- and trimers of butadiene	90 %	85 %	78 %
XXIIIa	2.9%	4.6%	7.3%
XXXII	1.0%	1.5%	1.5%
XXIIIc	0.2%	0.4%	2.8%
XVIa	2.4%	3.7%	5.5%
XVIb	1.3%	2.0%	1.3%
Higher oligomers	1.7%	3.0%	3.5%

From these data the following conclusions can be drawn:

1. Mainly cyclic butadiene dimers are formed on a nickel-triphenylphosphite catalyst.

2. The yield of C_{11} hydrocarbons depends on the molar butadiene-propylene ratio; 1 : 1 gives 7.8% and 1 : 2 gives 12.2% of co-trimers.

3. Using a low-valent nickel catalyst, which does not contain a phosphorus ligand, 18.4% of butadiene-propylene co-trimers are formed.

The structures and configurations of the co-trimers formed are shown in chapter 3.2. Figures 5 and 6. One open-chain product not having the *trans*-deca-1,4,9-triene skeleton was detected. On hydrogenation of XXXII undecane resulted.

6.2.6. CATALYTIC CO-TRIMERIZATION OF BUTADIENE AND ALLENE OR BUTADIENE
AND METHYLENE-CYCLOPROPANE

Two further co-trimerizations leading to derivatives of *cis,trans*-cyclodeca-1,5-dienes will be mentioned. When a 10 : 1 mixture of butadiene and allene

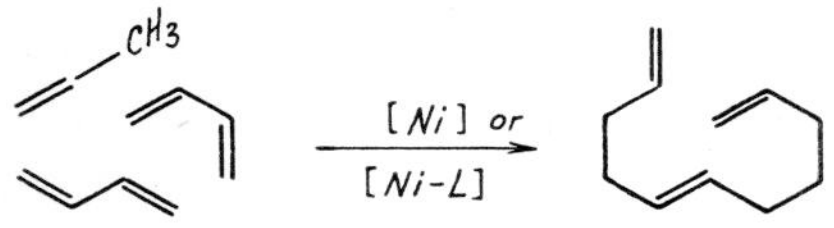

XXXII

(37)

is reacted at 40° C, two methylene-*cis*,*trans*-cyclodeca-1,5-dienes can be isolated in about 70% yield [41, 49].

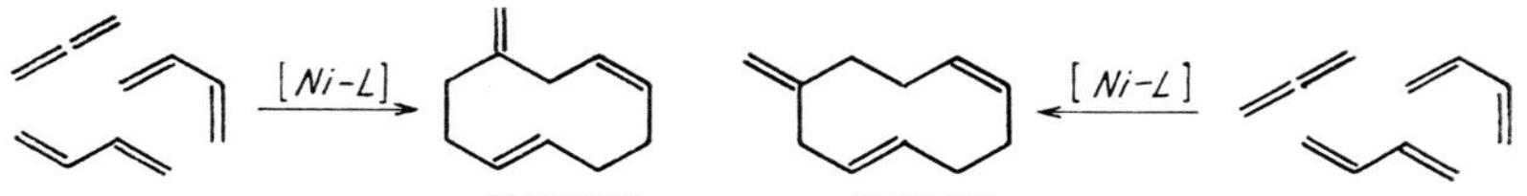

XXXIIIb XXXIIIa (38)

Both isomers thermally rearrange to the corresponding six-membered ring. It is interesting to note that the isomer XXXIIIa is a crystalline compound at lower temperature.

Therefore this isomer could be purified by recrystallization from ethanol. Its structure could be assigned by ^{1}H-NMR spectroscopy [41]. Partial hydrogenation (30% conversion) gives 8-methyl,*cis*,*trans*-cyclodeca-1,5-diene (XVIa) and other isomers:

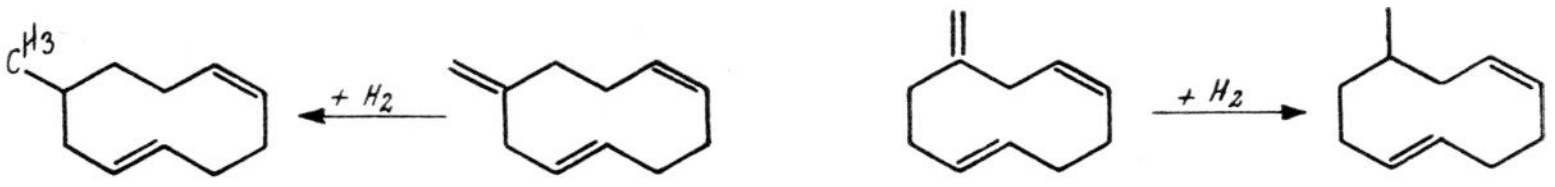

XVIa XXXIIIa XXXIIIb XVIb (39)

A mixture of XXXIIIa and XXXIIIb yields a mixture of XVIa and XVIb. Three different twelve-membered rings, formed from two allene- and two butadiene molecules, were detected as byproducts in the catalytic reaction. They were identified after hydrogenation to the 1,2-, 1,3- and 1,4-dimethylcyclododecanes (see chapter 4.4.).

Using a similar procedure to that shown in equation 39, we could determine the structures of two isomers formed catalytically from butadiene and methylene-cyclopropane [24, 50]. Instead of hydrogenation, XXXIIIa or a mixture of XXXIIIa and XXXIIIb was reacted with 1-3 mole-percent of diazomethane in the presence of copper powder:

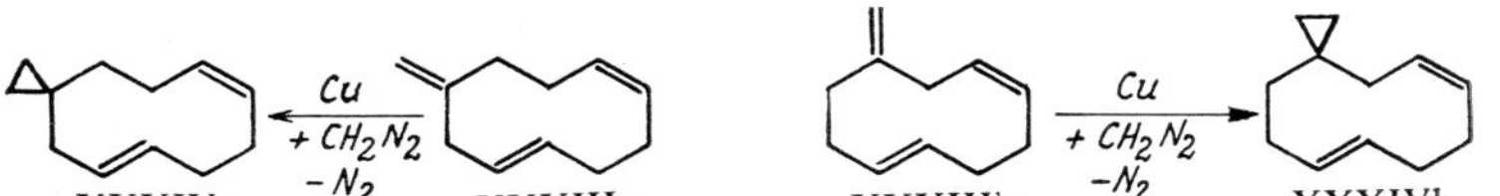

XXXIVa XXXIIIa XXXIIIb XXXIVb (40)

As well as two other cyclopropane-derivatives, not shown in equation 40, XXXVa or a mixture of XXXVa and XXXVb result (see chapter 3.).

In the co-trimerization of butadiene and methylene-cyclopropane (5 : 1) a third main product is formed on a nickel-triphenylphosphite catalyst:

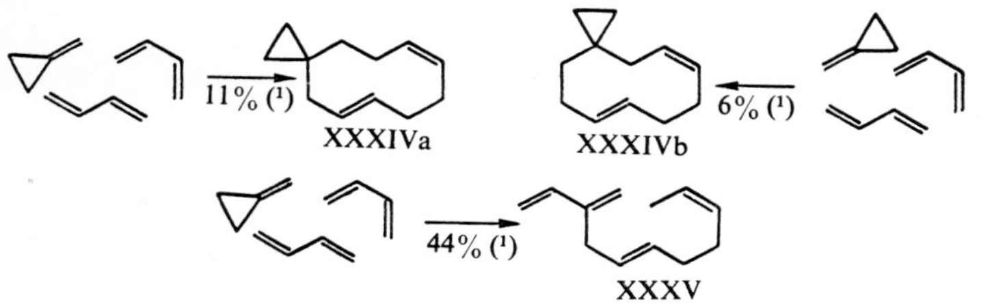

(41)

6.3. Catalytic syntheses of methyl substituted cyclododeca-1,5,9-trienes

The oligomerization of isoprene, piperylene and 2,3-dimethyl-butadiene with titanium-, chromium- and nickel-containing catalysts had been investigated by H. J. Kaminsky and G. Wilke [51, 52]. The cross-cyclotrimerizations of butadiene and isoprene or butadiene and 2,3-dimethylbutadiene were carried out with the same catalysts [41, 42]. Our intention was not to find more efficient catalysts than those described in the literature [1], but to prepare mono- and dimethyl-substituted cyclododeca-1,5,9-trienes.

In connection with questions discussed in chapter 8, we became interested in the configurations and structures of the rings formed and in the relative reaction rates of various 1,3-dienes promoted by various catalysts. In all experiments the reaction temperature was 40 °C.

6.3.1. CROSS-OLIGOMERIZATION OF BUTADIENE AND ISOPRENE

A 1 : 1-mixture of butadiene and isoprene was reacted with the catalysts in a glass-autoclave and the reaction was stopped before all 1,3-dienes were converted (control by volume contraction). In Table 12 are shown the experimental data obtained with various catalysts [41, 42].

The most remarkable fact concerning the relative reaction rates, is the difference observed using a nickel or a titanium catalyst. It is noteworthy that, in contrast to the results obtained with a nickel or titanium catalyst, only traces of dimers and cross-dimers (low boiling products) are found using the chromium catalyst. In the low boiling fraction of the reaction mixture (21%) obtained with the nickel catalyst, the following products were identified, ca. 20 compounds remaining unidentified:

4-vinylcyclohexene	2 0%
1-methyl-4-vinyl-cyclohexene (XXXVII)	2.2%
cis,cis-cycloocta-1,5-diene	1.2%
1-methyl-cis,cis-cycloocta-1,5-diene	0.8%

(*) Based on converted methylene-cyclopropane

Table 12

THE PRODUCT DISTRIBUTION AND RATES OF THE CROSS-OLIGOMERIZATION OF BUTADIENE
AND ISOPRENE WITH VARIOUS CATALYSTS [41]

Catalyst	Ni	Cr	Ti	
Low boiling products	30.0%	1.6%	41.0%	
CDT isomers	27.0%	5.1%	8.5%	
Me-CDT isomers	25.0%	23.0%	20.0%	
DMe-CDT isomers	3.9%	17.0%	12.0%	
Higher oligomers	14.0%	53.0%	18.0%	
Cross-trimer. of	0.11	8.8	27.1	mmole butad./mmole cat. · h
butad.-isoprene	0.04	5.8	27.3	mmole isopr./mmole cat. · h
$\left[\begin{array}{l}\text{cyclotrimeris.}\\\text{of butadiene}\end{array}\right]$	(2)	(120)	(380)	mmole butad./mmole cat. · h

In the low-boiling fraction obtained with the titanium catalyst, the following compounds could be identified:

1-methyl-4-vinyl-cyclohexene (XXXVII)	25.0%
p-diprene	7.7%
2,6-dimethyl-octa-1,*trans*-3,6-triene (XXXVIII)	6.2%

XXXVII was hydrogenated to *cis*-and *trans*-1,4-ethyl-methyl-cyclohexane. That the substituents were in 1,4-positions was proved by MIR of ethyl-cyclohexane. XXXVIII was hydrogenated to 2,6-dimethyl-octane: the structure was ascertained by H. J. Kaminsky [51].

Two open-chain cross-trimers were also formed with the nickel catalyst. After hydrogenation these could be identified as 3-methyl-dodecane (3.6%) and 2,10-dimethyl-dodecane (5,4% of the reaction mixture, included in table 12 under higher oligomers).

Better yields of 1-methyl-cyclododeca-1,5,9-trienes were obtained when the chromium catalyst was dissolved in a 1 : 1-butadiene-isoprene mixture, the further butadiene being injected into the autoclave via a magnetic valve as it was consumed. The yield of the 1-methyl-cyclododeca-1,5,9-trienes with respect to all products is greater than 30% while the higher oligomers are formed in less than 40% yield.

Tables 13 and 14 show the isomer distribution of the cyclododeca-1,5,9-trienes and that of the monomethyl derivatives, respectively. In Table 13 the

Table 13

THE ISOMER DISTRIBUTION OF THE VARIOUS CYCLODODECA-1,5,9-TRIENES FORMED IN THE CATALYTIC CROSS-OLIGOMERIZATION OF BUTADIENE AND ISOPRENE. IN ADDITION THE ISOMER DISTRIBUTION OF THE BUTADIENE REACTION UNDER SIMILAR CONDITIONS IS GIVEN IN PARENTHESES [41]

Catalyst	Ni	Cr	Ti
ttt-CDT (IVa)	90% (83%)	50% (59%)	40% (5%)
ttc-CDT (IVb)	7% (6%)	48% (38%)	60% (95%)
tcc-CDT (IVc)	3% (11%)	2% (2%)	— —

Table 14

THE ISOMER DISTRIBUTION OF THE MONOMETHYL-CYCLODODECA-1,5,9-TRIENES FORMED IN THE CATALYTIC CROSS-OLIGOMERIZATION OF BUTADIENE AND ISOPRENE [41]

Catalyst		Ni	Cr	Ti
1-Me-ttt-CDT	(XXVa)	85%	59%	38%
1-Me-ctt-CDT	(XXVb)	5%	12%	14%
1-Me-ttc + tct-CDT	(XXVc + d)	2%	29% (*)	48% (**)
1-Me-ctc + cct-CDT		5%	—	—
1-Me-tcc-CDT		3%	—	—

(*) 95% tct-CDT and 5% ttc-CDT.
(**) only ttc-CDT.

percentage distribution of the twelve-membered rings from the catalytic butadiene cyclotrimerization (without additional isoprene) is added in parentheses. The catalytically formed dimethylsubstituted cyclododeca-1,5,9-trienes show nearly the same isomeric distribution as found in the case of the monomethyl derivatives.

6.3.2. CATALYTIC-CROSS-TRIMERIZATION OF BUTADIENE AND 2,3-DIMETHYL-BUTA-DIENE

The same catalysts and experimental conditions were used as in the cross-trimerisation of butadiene and isoprene. The experimental data are summarized in Table 15.

All byproducts in the cross-cyclotrimerization of butadiene and 2,3-dimethyl-butadiene have analogous structures to those assigned to the byproducts from the cross-oligomerization of butadiene and isoprene. The isomeric distribution of the cyclic cross-trimers from 2,3-dimethyl-butadiene and two molecules of butadiene is shown in Table 16.

Table 15

THE PRODUCT DISTRIBUTION AND PERCENTAGE CONVERSIONS OF THE 1,3-DIENES IN THE CROSS-OLIGOMERIZATION OF BUTADIENE AND 2,3-DIMETHYL-BUTADIENE WITH VARIOUS CATALYSTS AT 40 °C [41]

Catalyst	Ni	Cr	Ti
Σ?	2.0%	1.0%	3.0%
Cyclic di- and trimers of butadiene	85.0%	16.0%	11.0%
Cyclic and open-chain cross-dimers	0.9%	—	27.0%
Open-chain cross-trimers	1.2%	14.0%	—
Σ 1,2-DMe-CDT	4.9%	13.0%	20.0%
Higher oligomers	6.0%	56.0%	39.0%
Conversion: butadiene	58 %	34 %	77 %
Conversion: 2,3-dimethyl-butadiene	6 %	24 %	56 %

Table 16

THE ISOMER DISTRIBUTION OF THE DIMETHYL-CYCLODODECA-1,5,9-TRIENES FORMED IN THE CATALYTIC CROSS-CYCLOTRIMERIZATION OF BUTADIENE AND 2,3-DIMETHYL-BUTADIENE [41]

Catalyst	Ni	Cr	Ti
1,2-DMe-ttt-CDT	55%	27%	79%
1,2-DMe-ctt-CDT	31%	15%	9%
1,2-DMe-ttc-CDT	5%	14%	8%
1,2-DMe-ctc-CDT	9%	44%	4%

7. MECHANISM OF CARBON-CARBON BOND FORMATION PROMOTED BY LOW VALENT NICKEL CATALYSTS

In this chapter our present ideas concerning the mechanism of carbon-carbon bond formation promoted by low-valent nickel catalysts will be reported [53]. In certain cases these differ from our previous formulations of the stereochemistry of the intermediate complexes. Our aim is that of discussing these new ideas in conjunction with some general aspects of carbon-carbon bond formation reported in chapter 8.

It was always assumed [5, 6, 7] that an eight-membered chain, formed from two molecules of butadiene, plays a significant rôle in the cyclodi- and trimerization, as well as co-oligomerization, reactions carried out with the help of low-valent nickel catalysts.

A single interpretation of all reactions can be given, if one assumes the formation of a *trans*-π-allyl-*cis*-σ-allyl-C$_8$-chain:

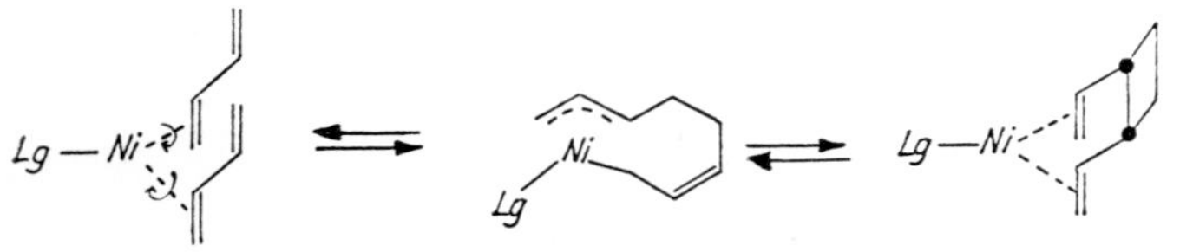

$$(42)$$

XXXVIII   XXXIX   XL

All three complexes XXXVIII, XXXIX and XL, shown in equation 42, can be isolated and their structures ascertained by ¹H-NMR spectroscopy [54, 55] (see also chapter 8.2.2.). It is assumed that the eight-membered chain is formed by "reductive activation" (see chapter 8.2.4.) of the 1,3-diene or the 1,2-divinyl-cyclobutane. The following reaction may support this assumption: H. Hey reacted *cis*-1,2-divinyl-cyclobutane with sodium in liquid ammonia and found that a stereospecific ring-opening occurred [56]:

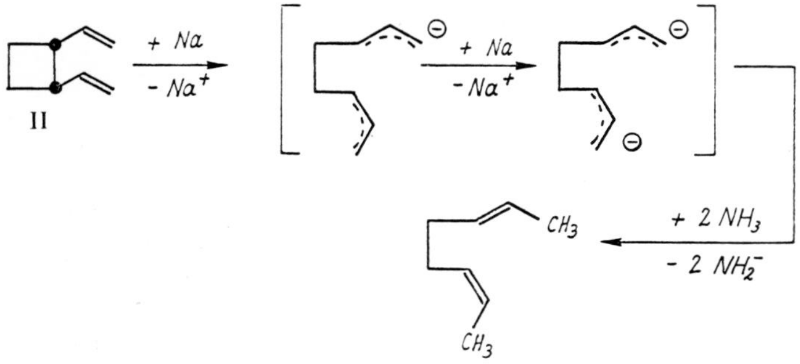

$$(43)$$

The formation of *cis,trans*-2,6-octadiene may be an orbital-symmetry controlled process, ad is probably the formation of the eight-membered chain with a *cis,trans*-configuration in the complex (see chapter 8.2.5). As one can see from molecular models, two different forms of this complex may exist. An equilibrium between these two forms cannot be ruled out:

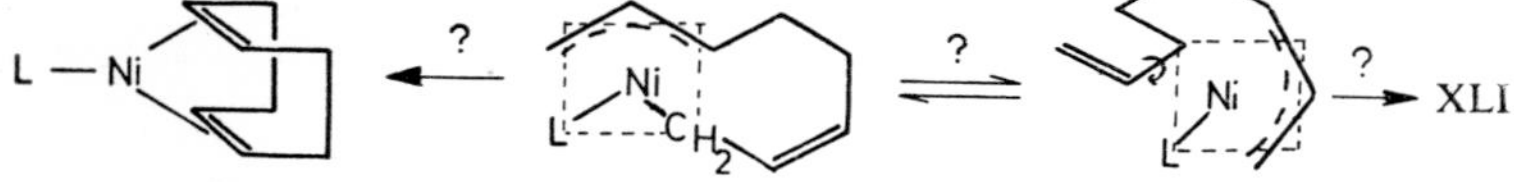

$$(44)$$

XLI   XXXIX   XLII

As shown by the model, the square-planar complex XLII, which has a *cis-π*-allyl-3-hapto-allyl-chain (*), is strained to some extent. However, one can assume that *cis,cis*-cycloocta-1,5-diene is formed from this arrangement. It is nevertheless possible, that the *cis,cis*-cycloocta-1,5-diene may be formed directly from the complex XXXIX, with a similar activation energy than *cis*-1,2-divinyl-cyclobutane.

Furthermore all reactions of the C$_8$-chain of XXXIX, are:

a) with further ethylene to give *trans*-deca-1,4,9-triene and *cis,trans*-cyclodeca-1,5-diene,

b) with further but-2-yne to give 4,5-dimethyl-*cis,cis,trans*-cyclodeca-1,4,7-triene,

c) with secondary amines to give *trans,trans*-octa-1,3,6-triene and *cis,trans*-octa-1,3,6-triene

d) with further butadiene to give a twelve-membered chain with an internal *trans* double bond (see also chapter 8.2.4).

All these can be interpreted as reactions of *trans-π*-allyl groups (see Figure 21).

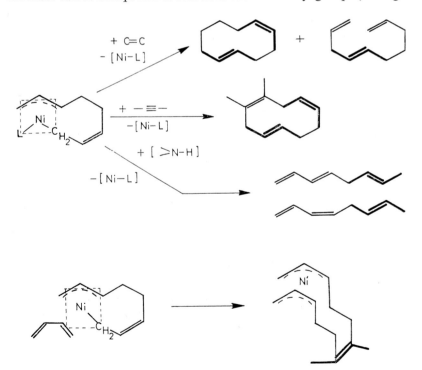

Figure 21. Various reactions of the *trans-π*-allyl-1-hapto-allyl-C$_8$-chain XXXIX at a nickel-ligand catalyst.

(*) For hapto nomenclature see [57].

We assume, that these are nucleophilic exchange reactions; the nichel-atom remains in the formal oxidation-state two.

In the further reactions of the complex XXXIX the fifth co-ordination position on the nickel may be occupied by the reactants (ethylene, alkynes or secondary amines). This, presumably tetragonal-pyramidal, complex XLIII may rearrange with carbon-hydrogen or carbon-carbon bond formation:

XXXIX 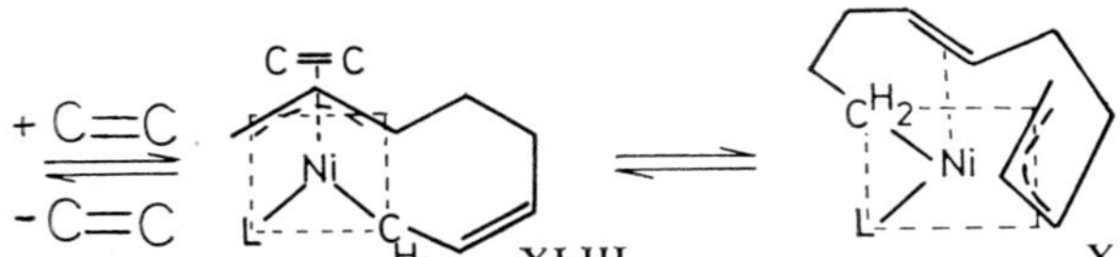 (45)

XLIII XLIV

XXXIX 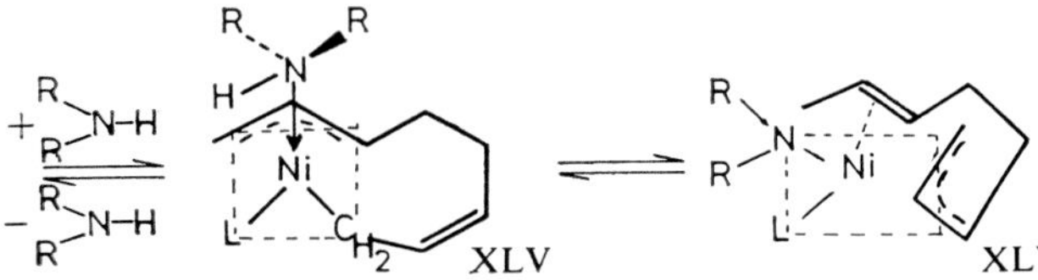 (46)

XLV XLVI

For example, after the rearrangement of complex XLIII a new complex XLIV may be formed, which has a π-allyl-group, a metal-carbon-σ-bond, an (albeit weak) metal-olefin interaction in the apical position and a nickel-ligand bond as in complex XLIII. In the case of complex XLV, the rearrangement differs only in that a metal-nitrogen σ-bond is formed instead of a metal-carbon σ-bond. The last step in the formation of *trans,trans*- and *cis,trans*-octa-1,3,6-triene [45] is presumably the formation of the *s-cis*-1,3-dienes:

XLVI 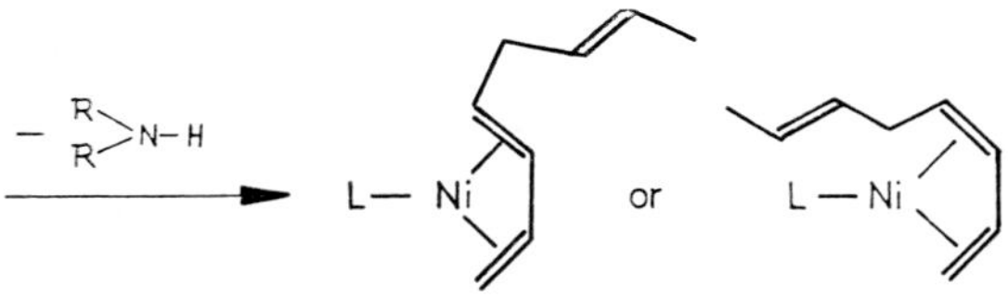 (47)

XLVII XLVIII

In the further reactions of complex XLIV the formation of a Ni(II)- or a Ni(0)- complex determines whether open-chain or cyclic co-trimers are formed (see later, Table 17 and the discussion of this experimental data).

As was mentioned in the introduction, a large variety of mono- and dimethylsubstituted ring olefins was prepared in order to investigate the influence of the methyl group on, for example, the properties of the intermediate complexes. The reactivities of various methyl substituted olefinic systems in catalysis can sometimes be interpreted by simple Hückel-MO considerations [58]. In the next few pages the formation and reactivity of various methyl-substituted C_8-chains will be discussed.

Table 17

THE DEPENDENCE OF THE PRODUCT DISTRIBUTION IN THE CO-TRIMERIZATION OF ISOPRENE AND ETHYLENE ON THE NATURE OF THE LIGAND [24]

Ligand	$P(C_6H_{11})_3$	$P(OC_2H_5)_3$ (*)	$P(C_6H_5)_3$	$P(OC_6H_5)_3$	$P(OC_6H_4C_6H_5)_3$
(first Ni complex) product A	96%	94%	90%	79%	14%
(ratio)	9.6%	20.2%	15.7%	13.3%	2.7%
product B	4%	6%	10%	21%	86%
(second Ni complex) product A	100%	99%	99%	97%	80%
(ratio)	19.1%	27.1%	20.9%	28.3%	52.9%
product B	—	1%	1%	3%	20%

Ligand	P(C₆H₁₁)₃	P(OC₂H₅)₃ (*)	P(C₆H₅)₃	P(OC₆H₅)₃	P(O–C₆H₄–C₆H₅)₃
(Ni complex 1) → products	71% / 29%; 64.3%	55% / 45%; 47.7%	31% / 69%; 54.0%	17% / 83%; 52.5%	3% / 97%; 39.9%
(Ni complex 2) → products	100% / —; 7.0%	100% / —; 5.0%	100% / —; 9.4%	100% / —; 5.9%	100% / —; 4.5%

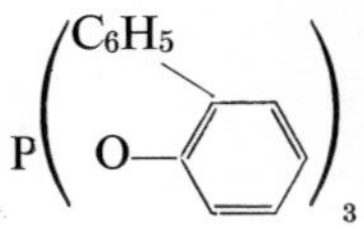

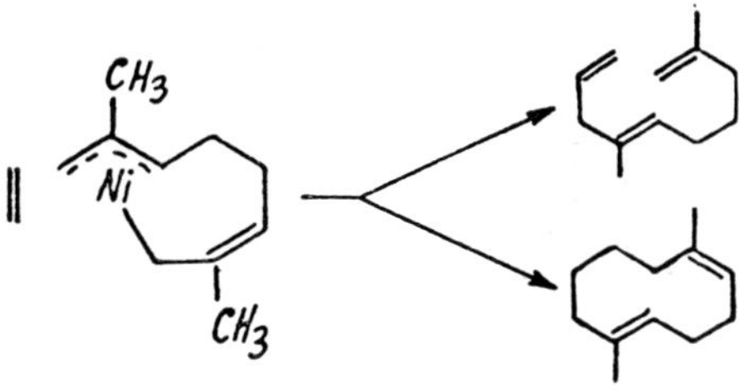

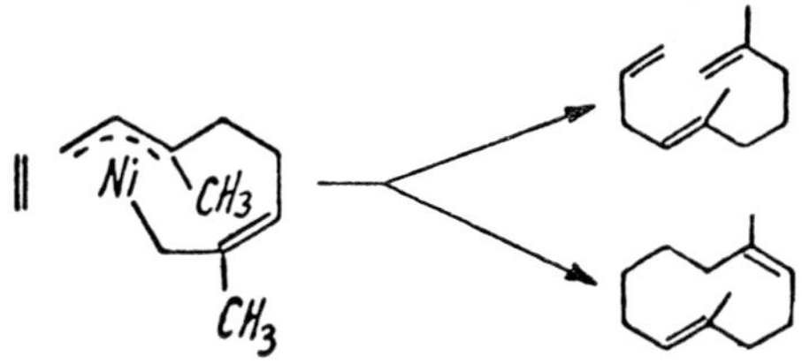

(*) In this case we have to assume a ligand-disproportionation (see Figure 24 and equation 53).

136

Two eight-membered chains are formed from the reaction of two piperylene molecules using the nickel-ligand catalyst (head to tail addition in complex L and tail to tail addition in complex IL; see also chapter 6.2. equations 28 and 29):

(48)

(49)

The relative rates of formation of complex IL and complex L are as yet unknown, as the indirect determination of these rates from the reaction products is not possible. The four-membered rings VIIa and VIId (see chapter 6.1. Figure 17), obtained from complex IL and L (L = triphenylphosphine), are formed initially in the ratio 93 : 7, whereas the ten-membered rings XVIIIa and XVIIIc from *trans*-piperylene and ethylene (see chapter 6.2. Table 8) are formed in the ratio 8 : 92. The mode of addition (head to tail or tail to tail) depends on several factors, including the ligands on the nickel. For instance in the co-cyclo-trimerization of *cis*-piperylene and ethylene 14% head to tail and 86% tail to tail products were isolated (in the case of *trans*-piperylene and ethylene this ratio was 92 : 8). Head to head addition of two piperylenes was not found, except in the co-oligomerization of piperylene and ethylene with a phosphane or phospite free nickel catalyst. In this case a twelve-membered ring, arising from two ethylene and two head to head linked piperylene molecules, could be detected. We assume in this case that firstly two ethylene molecules react with the catalyst and then further reactions with piperylene occur (see also chapter 8.2.3.) [24]:

(50)

(*) Olefinic ligand

Four isomeric C_8-chains can be synthesized from two isoprene molecules (see Table 17). Further reactions of these complexes with ethylene give four different open-chain and three cyclic-co-trimers in high yields. We investigated the dependence of the product distribution of this co-trimerization on the nature of the ligand. The experimental data is summarized in chapter 6.2. Table 7. Taking the sum of co-trimers as 100%, Table 17 shows the percentage distribution of the individual open-chain and cyclic products (underlined). The additional percentages in Table 17 give the proportion of open-chain to cyclic isomers.

In two publications [59, 60] C. A. Tolman succeeded in grading the phosphites and phosphanes according to their electronic and steric properties.

From these papers one can obtain an order of steric and electronic effects of those ligands, used in the co-trimerization of isoprene and ethylene (see Table 18).

Table 18

THE STERIC AND ELECTRONIC EFFECT OF SOME PHOSPHITES AND PHOSPHINES [59, 60].

Steric order	Minimum cone-angle (*) (degrees)	Electronic order	ν_{CO} (A$_1$)
P(OC$_2$H$_5$)$_3$	107° ± 2°	P(C$_6$H$_{11}$)$_3$	2056.4 cm^{-1}
P(OC$_6$H$_5$)$_3$	121° ± 10°	P(C$_6$H$_5$)$_3$	2068.9 cm^{-1}
P(C$_6$H$_5$)$_3$	145° ± 2°	P(OC$_2$H$_5$)$_3$	2076.3 cm^{-1}
P(C$_6$H$_{11}$)$_3$	179° ± 10°	P(OC$_6$H$_5$)$_3$	2085.3 cm^{-1}

Tri(o-phenylphenyl)phosphite presumably has a greater steric effect than tricyclohexylphosphane and an electronic effect similar to triphenylphosphite.

As can be seen from tables 17 and 18, steric effects hardly influence the product distribution. In this case one has to consider that only one coordination position of the nickel is occupied by a phosphorus containing ligand. The relative ratio of open-chain to cyclic co-trimers is mainly affected by the relative donor--acceptor properties of the ligands. The proportion of open-chain products formed depends on the relative stabilization of Ni(II) in complex XLIV of equation 45. The open chain isomers are formed by β-elimination in the alkyl group with transfer of the hydrogen to the cis-π-allyl-group and formation of a terminal double bond. This interpretation is confirmed by results obtained by H. Bönnemann [61]. The following order of stability was found investigating 0.5 molar solutions of π-allyl-nickel-ethyl complexes:

L = tri(o-phenylphenyl)phosphite: decomp. temperature   — 40 °C

L = triphenylphosphane: decomp. temperature   — 60 °C

L = tricyclohexylphosphane: complex not isolable at   — 130 °C

Increasing acceptor ability of the ligand leads to increasing amounts of cyclic products (see Table 17).

(*) The ligand cone angles were based upon a model which assumed tetrahedral bonding for phosphorus and a metal-phosphorus bond length of 2.28 Å [60].

Two different eight-membered chains are formed in the head to tail addition of two isoprene molecules:

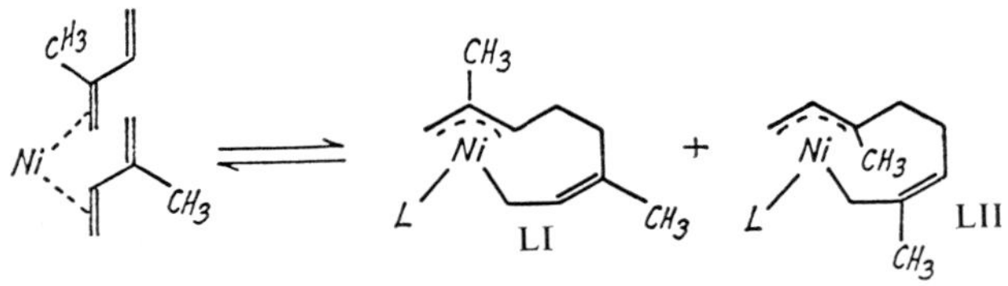

(51)

The intermediate complexes LI and LII are always formed or react with a further ethylene in the ratio 90 : 10. The following additional rules were found (see complex LIII):

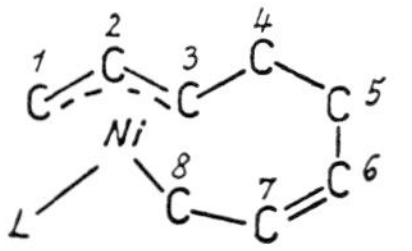

1. Mainly C_8-chains, having a methyl group at the C_2-atom, are formed (or react with further ethylene).

2. Those eight-membered chains, in which the second methyl group is bonded to the C_6-atom, give predominantly cyclic products.

In the case of the cyclic isomer XIIc, which is not formed catalytically, neither of these rules is fulfilled. The second rule indicates that the cyclic co-trimers possibily arise via an alkyl-l-hapto-allyl complex LIV leading by C-C bond formation to a zerovalent Ni-complex.

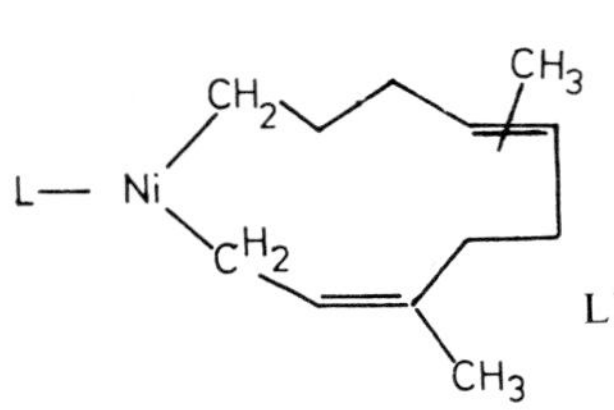

Some interesting results can be obtained when the eight-membered chains result from two different 1,3-dienes.

The open-chain and cyclic cross-cotrimers from butadiene, isoprene and ethylene can be formed via the following C_8-chains (see chapter 3.2. Figure 5 and 6 and chapter 6.2. Table 9):

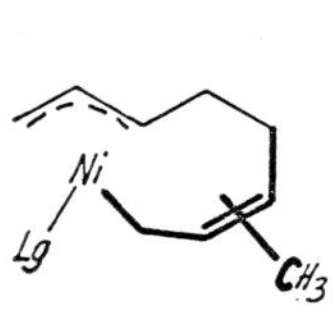

LV   or   LVI

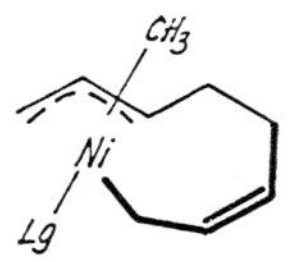

The same intermediates are assumed to occur in the cross-dimerization of butadiene and isoprene to monomethyl substituted *trans,trans*- and *cis,trans*-octa-1,3,6-trienes promoted by amine modified nickel-ligand catalysts (see chapter 5 Figures 14 and 15). From the product distribution we know that the complexes are formed (or react) in the cross-co-trimerization in the ratio 77 : 23 [24]. In the formation of open-chain cross-dimers this ratio is 84 : 16 [45]. In the cross-cotrimerization of butadiene, piperylene and ethylene (see chapter 3.2. Figures 5 and 6 and chapter 6.2. Table 10) no open-chain and cyclic isomers, which might be formed via the intermediate-complex LVII, could be detected:

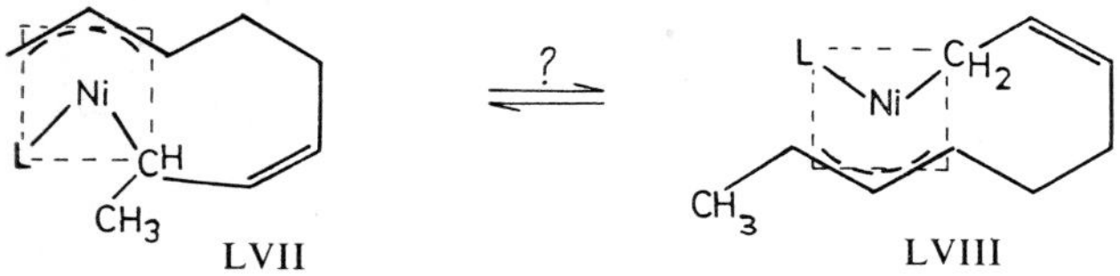

LVII   LVIII   (52)

We therefore assume that also in the co-trimerization of piperylene and ethylene intermediate complexes such as:

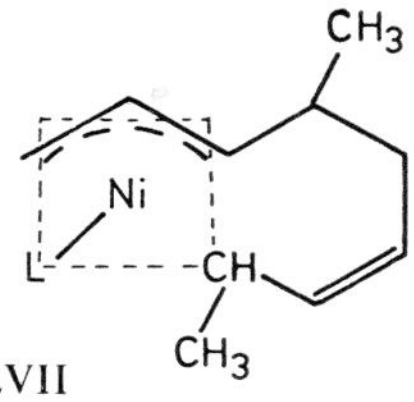

LVII

do not react (see chapter 2.3. equations 14 and 15 and chapter 6.2.). It is a matter of debate as to whether the complex LVII is not formed or whether it immediately rearranges to LVIII. This unusual rearrangement within the complex would form a *trans*-π-allyl- from a *cis*-1-hapto-allyl-group and at the same time a *cis*-1-hapto-allyl- from a *trans*-π-allyl-group.

In the co-oligomerizations of butadiene with propylene, allene or methylenecyclopropane (see chapter 6.2.) we also assume that only a *trans*-π-allyl-group reacts with the mono-olefins. The carbon-carbon bond formation preferentially occurs at the C_2-atom of the olefin:

$$
\begin{array}{ccc}
\mathrm{CH_3} & \mathrm{CH_2} & \mathrm{CH_2} \\
| & \| & \triangle \\
\mathrm{CH} & \mathrm{C} & \mathrm{C} \\
\| & | & | \\
\mathrm{CH_2} & \mathrm{CH_2} & \mathrm{CH_2}
\end{array}
\qquad \longleftarrow \text{ preferred reactivity}
$$

That means that the olefins react as nucleophiles with the metal.

8. SOME GENERAL ASPECTS OF CARBON-CARBON AND CARBON-HYDROGEN BOND FORMATION PROMOTED BY TRANSITION METALS

We are not really convinced that the intensive investigations in nickel organic chemistry will lead to a complete understanding of catalytic carbon-hydrogen and carbon-carbon bond formation promoted by all the transition metals. In this chapter, however, some general aspects of catalysis will be discussed as a conclusion of the many experimental data reported in this Review.

This discussion will deal with few connected aspects of homogeneous catalysis.

In all our publications on nickel and other transition metal catalysts a step-wise reaction mechanism was considered [4, 7, 18, 24, 45, 53] and in the interpretation of the catalytic reactions a general scheme of β-elimination and -addition reactions was very helpful [62].

8.1. General scheme of β-elimination and addition reactions

In the first part of the general scheme in Figure 22 is shown the β-elimination and -addition reactions of hydrogen and olefinic substrates aided by metal-catalysts. It must be emphasized that a scheme and not a series of equations is being discussed. For the sake of clarity only carbon atoms bonded to the metal are shown. It is obvious that the reaction of an alkyl complex to give an olefin metal hydride complex will only occur if the alkyl group contains at least two carbon atoms and if there is a hydrogen attached to the β-carbon atom. Similarly, an equilibrium between a π-allyl and a σ-allyl system can only exist if the metal-alkyl complex has a double bond adjacent to the β-carbon atom.

Any equilibria between diene and corresponding monolefin complexes are neglected in order to simplify matters. The addition to, or the elimination of, hydrogen from the complex is shown without introducing any charged species, since the hydrogen may enter or leave the complex as a proton, hydride ion or even as a neutral H_2 molecule. The scope of the scheme shown in Figure 22 can be extended if one replaces the hydrogen by an alkyl, olefin, allyl or 1,3- diene ligand. This is shown in Figure 23, where all the possible reactions between al-

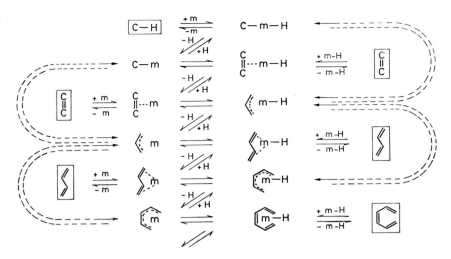

Figure 22. General scheme of carbon-hydrogen bond formation and fission by catalytic β-elimination and β-addition reactions. m = one or more equivalents of a main group or transition metal.

kyl, olefin, allyl and 1,3-diene ligands are summarized. The application and the combination of the reactions in these two schemes is quite general and is not limited to β-elimination and -addition reactions with nickel and other transition metal complexes. For example, the olefin chemistry of organo-aluminium compounds may be summarized by such a scheme. In this case, in view of the consi-

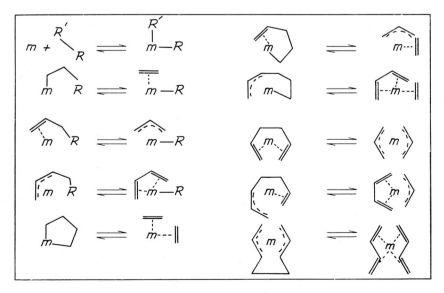

Figure 23. General scheme of carbon-carbon bond formation and fission by β-elimination and β-addition reactions. m = one or more equivalents of a main group or transition metal.

142

derably poorer π-bonding ability of the aluminium, only mono-olefin or at best dynamic allyl systems need to be considered.

The following main criteria are important in such a scheme:

1. Both reaction partners must interact with the catalyst (reactions from inside the catalyst).

2. The reactions may intuitively be supposed to occur with a minimum alteration in coordination number (rule of minimum coordination alteration, see also [111]).

3. Most of the β-elimination and -addition reactions occur in a stepwise fashion and are occasionally complicated by rearrangements of the intermediate complexes.

Some general aspects of the mechanism of carbon-carbon bond formation promoted by transition metals will be discussed with the help of a generalized potential energy diagram (chapter 8.2.).

Prior to this, however, I will briefly mention some problems concerning the concentration of the "active catalyst species". If the reaction is genuinely catalytic — and not stoichiometrically catalyzed — the "active species" must be regenerated in the course of a reaction cycle. In most cases the catalytic procedure involves a series of interdependent cycles and side equilibria. The metal

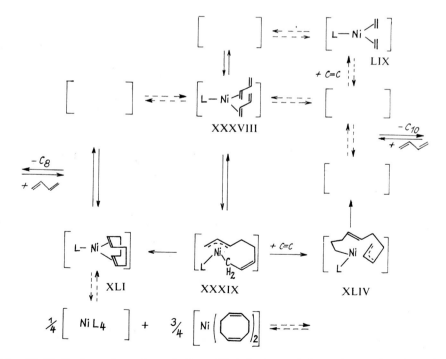

Figure 24. Idealized reaction cycles and side equilibria in the catalytic co-trimerization of butadiene and ethylene.

is mainly present in "inactive complexes". The latter will be discussed in context with the co-oligomerization of butadiene and ethylene with a nickel-triphenylphosphite catalyst [11]. The idealized reversible and irreversible reactions are shown in Figure 24.

To simplify matters, the further cycles, which lead to the formation of, e.g., 4-vinyl-cyclohexene, *cis*-1,2-divinylcyclobutane and cyclododeca-1,5,9-trienes, and a series of intermediate-complexes are omitted. The dotted arrows indicate that further intermediate complexes are perhaps involved. The concentration of ethylene determines whether more eight- or more ten-membered rings are formed. However, excessive increase in the ethylene concentration converts large amounts of calatyst to inactive complexes such as LIX in side-equilibria. Another decrease in "active species" can be caused by "ligand disproportionation" according to the following equations:

$$[NiL_4] \underset{-L}{\overset{+L}{\rightleftharpoons}} [NiL_3] \underset{-L}{\overset{+L}{\rightleftharpoons}} [NiL_2] \underset{-L}{\overset{+L}{\rightleftharpoons}} [NiL] \underset{-L}{\overset{+L}{\rightleftharpoons}} [Ni] \qquad (53)$$

Perhaps one advantage of the sterically demanding ligands, such as tri(*o*-phenylphenyl)phosphite, is that a tris- or tetrakis-complex cannot arise and almost all the ligand is used in the formation of a 1 : 1-complex (see [10]).

8.2. Discussion of a generalized potential energy diagram

The generalized potential energy diagram in Figure 25 shows which different possibilities have to be considered in discussing the pathways of catalytic

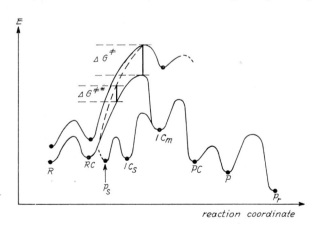

Figure 25. Generalized potential energy diagram

R = reactant	RC = reactant-complex
P = product	PC = product-complex
P_r = thermally rearranged product	
P_s = product in side equilibrium	
IC_m = intermediate complex in main path	
IC_s = intermediate complex in side equilibrium	

carbon-carbon bond formation with respect to the organic reactants and products. The nomenclature is adopted from [63]. In the case of a metal catalyzed carbon-carbon bond formation the following aspects will be discussed in some detail:

1. Structure and configuration of the catalytically formed products.
2. Preparation and isolation of the complexes.
3. Kinetic investigations and competitive reactions.
4. Redox processes and ligand interaction in catalysis.
5. Conservation of orbital symmetry in concerted, stepwise catalytic processes.

In the following pages are summarized some experimental details and also some speculations.

8.2.1. STRUCTURE AND CONFIGURATION OF THE CATALYTICALLY FORMED PRODUCTS

The cyclodimerizations of pure *cis*-, pure *trans*- or mixtures of *cis*- and *trans*-piperylene, promoted by nickel-ligand catalysts (see Chapter 6.1.), convincingly show that the exact determination of the configurations and structures of the catalytically formed products leads to a significant understanding of carbon-carbon bond formation. This, of course, is also valid for the various ten-membered ring syntheses (see chapter 7.) as in principle for all catalytic reactions.

A further aspect is to exclude possible rearrangements of the catalytically formed products during the work-up. *Cis*-1,2-divinylcyclobutanes (see chapter 2.2.) and *cis,cis,trans*-cyclodeca-1,4,7-trienes [23] rearrange on heating to 60 °C for instance. The catalytically formed 5-vinyl-cyclohexa-1,3-dienes (see chapter 8.2.3.) [64, 65] also rearrange during distillation:

(54)

In a recent publication the formation of tricyclo-$(2,2,2,0^{2,6})$oct-7-enes from two but-2-yne-molecules and butadiene promoted by a titanium-catalyst is described [66]. Using a homogeneous Ti-catalyst a 5-vinyl-cyclohexa-1,3-diene derivative is also the primary product [67].

From the reaction of butadiene with Reppe catalysts [3, 68] (further activated by acetylene) H. W. B. Reed was able to isolate small amounts of a compound which was suggested to be 4,5-divinyl-cyclohexene.

"Whatever the isomer, it is noteworthy that it is formed from two molecules of butadiene both reacting in the 1 : 2-position" [2]. Although it was not recognized, this was the first catalytic synthesis of a ten-membered ring. The *cis*,

cis,trans-cyclodeca-1,4,7-triene formed rearranges on distillation to *cis*-4,5-divinyl-cyclohexene [69]:

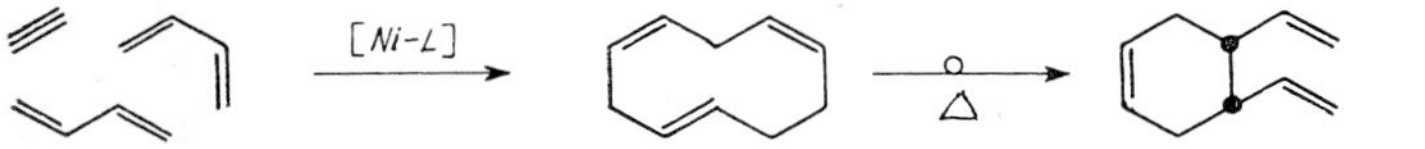

(55)

E. Koerner von Gustorf [70] has assumed that the rearrangements of highly strained hydrocarbons induced by transition metal catalysts sometimes involve a sequence of catalytic and thermal reactions. This is an interesting aspect and should be considered in formulating the mechanism of these reactions.

8.2.2. Preparation and isolation of the complexes

Various types of complexes may be prepared to aid the interpretations of the catalytic reactions:

reactant complexes and product complexes,

intermediate complexes and model complexes.

Only one or two examples of each type of these complexes will be discussed. In chapter 7, equation 42, is shown an example in which the reactant, the intermediate and the product complex could be isolated [54, 55]. As was discussed in this chapter, there is good experimental evidence, that the isolated intermediate complex is involved in the main reaction path.

Another intermediate complex, which could be isolated in the reaction of bis-cyclooctadiene-nickel with three molecules of butadiene, must participate as a bis-π-allyl-complex in a side-path of the reaction, as can be seen from the configurations of the products and their assumed formation, discussed in chapter 8.2.4:

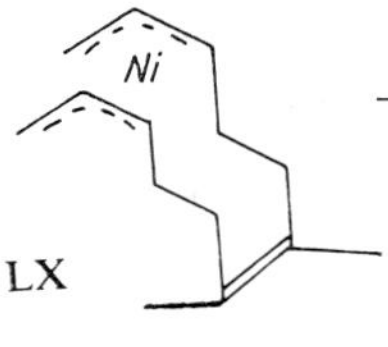

95% ttt-CDT, 5% ttc-CDT $\xleftarrow[-\mathrm{Ni[P(C_6H_5)_3]_4}]{+4\mathrm{P(C_6H_5)_3}}$ (20 °C)

LX

$\xrightarrow[-\mathrm{Ni(COD)}]{+2\mathrm{COD}}$ (80 °C) 48% ttt-CDT, 15% tcc-CDT, 37% ttc-CDT (56)

The configurations of the complex LX were ascertained by ^{1}H-NMR spectroscopy [9].

Some interesting model complexes for nickel catalysis were synthesized by G. Wilke and H. Boennemann [112]:

$$\mathrm{L{-}NiX_2} + \mathrm{Na[B(CH_3)_3H]} \xrightarrow[-\mathrm{B(CH_3)_3}]{-\mathrm{NaX}} \mathrm{L{-}Ni(H)X} \qquad (57)$$

The complex LXI and its reaction with propylene [71] was of interest concerning the propylene dimerization [72, 73].

H. Schenkluhn [74, 75] extensively investigated the interaction of various ligands with the relatively stable syn,syn-1,3-dimethyl-allyl-nickel-methyl complex LXII:

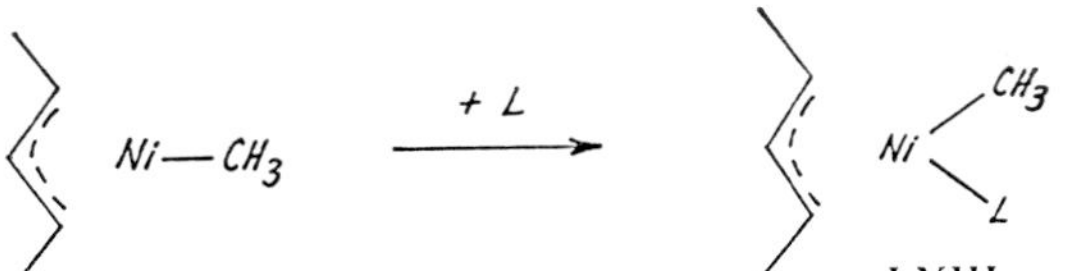

$$(58)$$

On interaction with phosphanes and phosphites a slow rearrangement was observed in the complex:

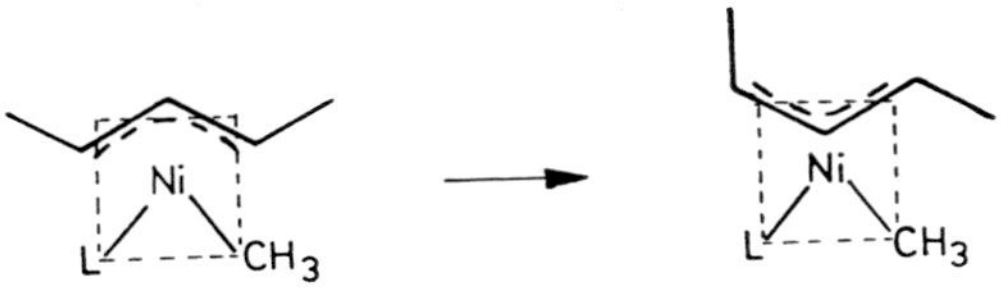

$$(59)$$

If a similar rearrangement is assumed to occur in the eight-membered chain, formed from e.g. two *trans*-piperylene molecules, and the C_8-chain is again split into 1,3-dienes, this affects an isomerization of one *trans*-piperylene molecule to the *cis*-isomer (see also chapter 6.1. equation 22).

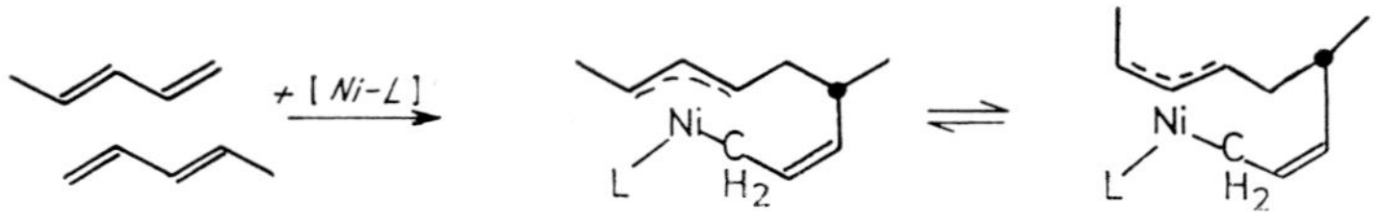

$$(60)$$

Other model complexes, in which the relative stability of complexes with variously substituted olefins can be obtained, were synthesized by K. Jonas and G. Wilke [76, 77, 78]:

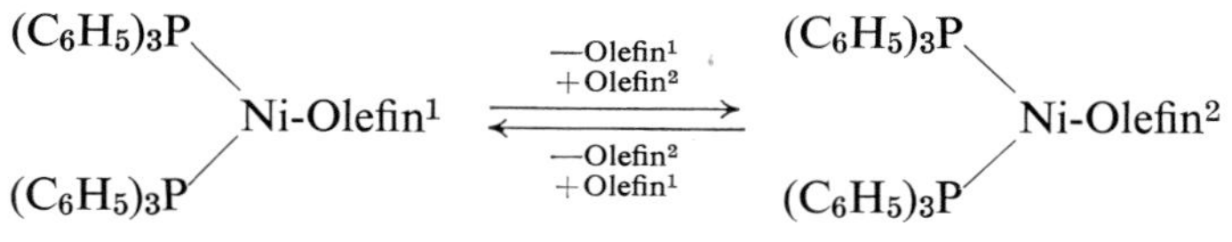

$$(61)$$

The order of stability, obtained from these qualitative experiments, is to some extent in agreement with results from competitive catalytic reactions (see next chapter).

8.2.3. Kinetic investigations and competitive reactions

As was shown in chapter 6.1. the concentration time curves can give an insight into the intermediate product distribution. From these experiments we could conclude that the *cis*-1,2-divinyl-cyclobutane is presumably formed in a side equilibrium.

I do not intend to discuss the inherent difficulties involved in kinetic measurements (see chapter 8.1. Figure 24). However, I will give two examples to show that a great deal of information can be obtained from simple competitive reactions.

We reacted butadiene and various substituted olefins with a phosphane and phosphite free nickel catalyst. The hydrogenated cyclotrimers of butadiene as well as the hydrogenated open-chain and cyclic co-trimers and co-oligomers of butadiene and these substituted olefins are summarized in Figure 26 [79, 80].

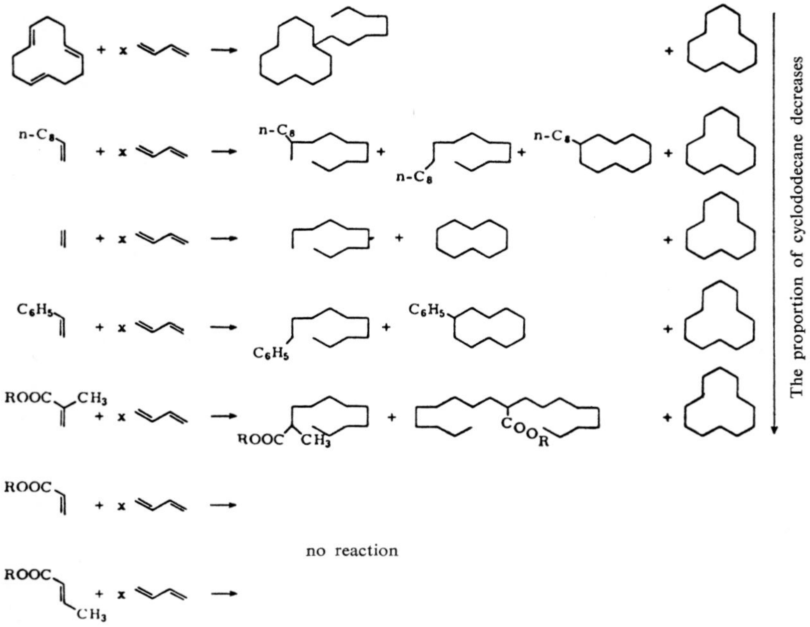

Figure 26. The qualitative dependence of the product distribution in the co-oligomerization of butadiene with substituted olefins on the type of substitution in the mono-olefins. The hydrogenated main products are listed. Catalyst: low-valent nickel without a phosphorus containing ligand.

The better the interaction of the olefin with the catalyst the less cyclotrimers of butadiene and the more co-oligomers are formed. If the olefin interacts too strongly with the transition-metal, no catalytic reaction occurs at all.

The order of reactivity of the different substituted olefins correlates with the order of stability of the bis-phosphine-nickel-olefin complexes:

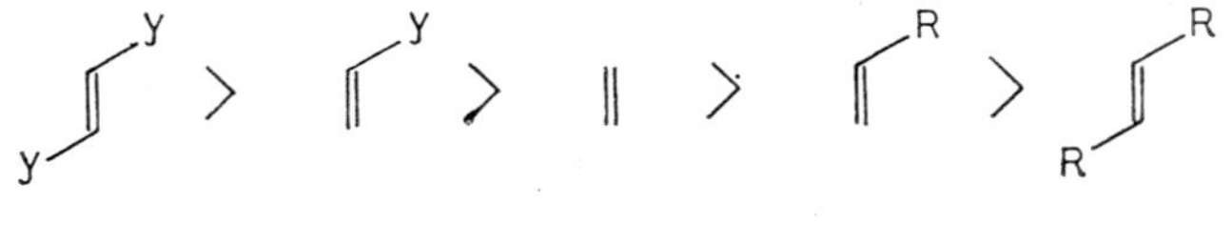

With olefins having an electronic push-pull effect (e.g. crotonic acid ester) discrepancies in these orders are found. This phenomenon may be explained by the fact that in the bis-trialkylphosphane nickel complexes the "trunk-complex" NiL_2 (i.e. the rest of the complex) is symmetric. However, if the only ligands are olefins which have an electronic push-pull effect, the complex becomes asymmetrically perturbed and stabilized by a series of trans-effects.

The second example that will be discussed is the co-oligomerization of butadiene with various alkynes. In these reactions the experimental results may be interpreted in terms of a competition between butadiene and alkyne molecules for the first carbon-carbon bond formation [13, 64, 65]. Whether two butadiene or two alkyne molecules react with each other in the first step depends on the substituents on the alkynes.

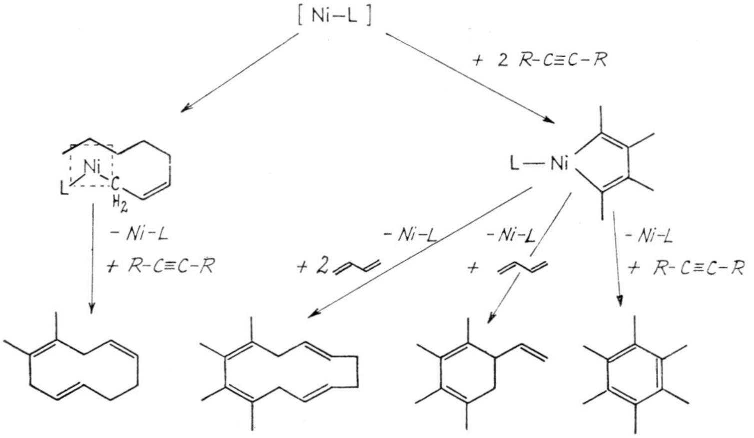

Figure 27. Scheme for the formation of various products in the catalytic co-oligomerization of butadiene and alkynes.

Using a nickel-triphenylphosphane catalyst, a mixture of butadiene and dialkyl-, alkylaryl- and diarylsubstituted alkynes results in the formation of 4,5-disubstituted *cis,cis,trans*-cyclodeca-2,4,7-trienes [12, 13, 14, 64, 82]. However, alkyl- and aryl-substituted acetylene-carboxylic esters form 5-vinyl-cyclohexyl-1,3-dienes with butadiene [64, 65]. Some byproducts of the co-oligomerisation of butadiene and alkynes are shown in Figure 27. In principle the twelve-membered rings, synthesized from two molecules of butadiene and alkyne, may arise via two different pathways:

1. Formation of the C_8-chain and subsequent reaction with two alkyne molecules.

2. Formation of a ligand-nickel cyclopentadiene complex and subsequent reaction with two molecules of butadiene.

We were able to isolate isomeric twelve-membered rings:

The *cis,cis,trans,trans*-isomers of type A possibly arise, as shown in Figure 27, via the second route whereas the *cis,cis,cis,trans*-isomers of type B may form via the first route (compare chapter 7. equation 50).

Finally it should be mentioned that competitive reactions carried out with model complexes and various substrates can also give useful information concerning catalysis.

8.2.4. REDOX PROCESSES AND LIGAND INTERACTION IN CATALYTICALLY ACTIVE COMPLEXES

Catalytic processes can readily be understood in terms of a series of redox processes. The question arises as to what individual processes have to occur in a metal complex in order that carbon-hydrogen or carbon-carbon bond fission and formation result. Oxidative addition and reductive elimination reactions, which are reviewed in [83, 84], are well understood:

$$[M(o)] \longrightarrow [M(II)] + 2e^- \tag{62}$$

Reactions of this type have been extensively investigated e.g. with Vaska's compound [85, 86]:

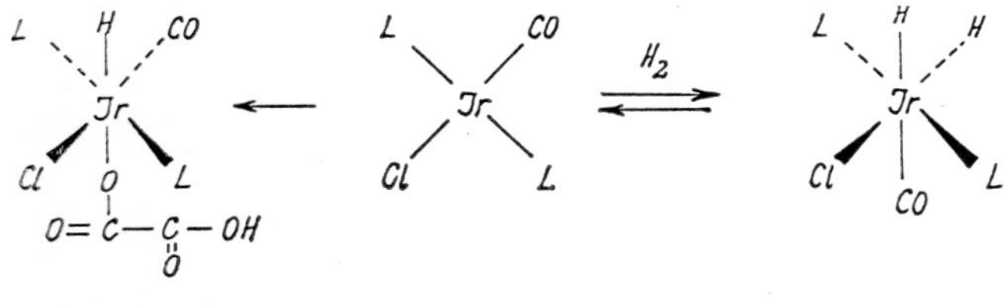

and with bis-ligand-nickel complexes [87, 88].

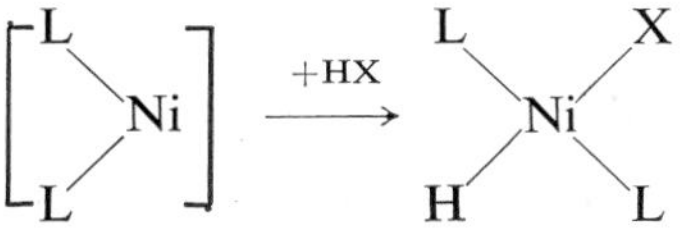

At this point it must be emphasized that these reactions occur with changes in the formal oxidation state and coordination number of the metal as well as with a change in the overall geometry of the complex. These changes are also assumed to occur in the formation of an eight-membered chain from two molecules of butadiene at a nickel-ligand catalyst (see chapter 7. and 8.2.5.). A formal redox process has already been assumed by G. Wilke [4]. On the other hand nucleophilic and electrophilic exchange reactions in the ligand-sphere of a metal are possible; in all these cases the metal remains in its oxidation state (see [113]).

Other processes, which may lead to bond formation or fission within the complex, involve the addition or elimination of a further ligand. This ligand will be referred to as the "accelerating ligand" [62]. Sometimes these reactions occur without change in the oxidation-state of the metal.

$$[M] \underset{-L}{\overset{+L}{\rightleftharpoons}} [L-M] \tag{65}$$

The accelerating ligand can have two effects:

1. A labile equilibrium between two complexes can easily be forced in one direction [89, 90].

$$R-Co(CO)_4 \rightleftharpoons R-\overset{\overset{\text{O}}{\|}}{C}-Co(CO)_3 \xrightarrow{+CO^*} R-\overset{\overset{\text{O}}{\|}}{C}-Co(CO)_3CO^* \tag{66}$$

2. Further complexation of an accelerating ligand can lead to an unstable complex, which can rearrange with bond formation (see for example the suggested mechanism in chapter 7. equation 45).

An equilibrium between monomeric and dimeric forms of a complex can also effect bond formation or fission:

$$2[L-M] \rightleftharpoons \left[M \overset{L}{\underset{L}{\diamond}} M \right] \tag{67}$$

Other processes, which can lead to bond formation or breaking, are summarized in the following equations (in these cases redox-processes of the metal may be facilitated):

$$\left[\overset{X}{\underset{X}{\diamond}} Al \overset{X}{\underset{X}{\diamond}} M \right] \rightleftharpoons \left[X-\overset{\overset{X}{|}}{\underset{X}{|}}{Al}-X-M \right] \tag{68}$$

The first process appears to play a significant role in the dimerization and isomerization [72, 73, 91] and polymerization [92] of olefins.

All these processes, and others not discussed here, lead to an activation of olefinic systems in the complex. In order to understand this activation one has to remember how, in principle, an unsaturated system may be activated. This will be discussed for butadiene. The 1,3-diene will react after nucleophilic or electrophilic attack. Oversimplifying this process, one may say that the reactivity is caused by the fact that two electrons will enter the ψ_3-orbital of the butadiene or will leave the ψ_2-orbital:

reductive   activation   oxidative

It was proposed [53] that the first process should be called a "reductive" and the second process an "oxidative activation". In a metal complex this activation normally involves an interaction of suitable filled d-orbitals of the metal with the empty ψ_3-orbital, or an interaction between the ψ_2-orbital of the 1,3-suitable empty d-orbitals. Another possibility is the reaction with nucleophiles or electrophiles of the ligand-sphere.

The change in the geometry of the metal complex (see chapter 7. equation 42 for instance), resulting from changes (e.g. rotation) in the interactions of the ligands with the metal or from addition or loss of an accelerating ligand, (which may be the reactant olefin itself) effects a change in the energy content and order of the d-orbitals of the metal. The necessary interactions between the ψ_2- or ψ_3-orbitals of butadiene and the suitable d-orbitals may arise in these processes.

A consequence of this concept is that butadiene, for instance, should be more easily reductively activated with nickel-ligand catalysts than isoprene or piperylene. This is indeed observed in e.g. the catalytic cyclodimerization of 1,3-dienes and the co-trimerization of 1,3-dienes with ethylene (see chapter 6.2. page [57]). On the other hand isoprene should be more easily oxidatively activated than butadiene and, as mentioned above, less easily activated in the case of reductive activation. The reaction rates of butadiene and isoprene in the cross-cyclotrimerization at a nickel (reductive activation) or a titanium catalyst (oxidative activation) follow this trend (see chapter 6.3. Table 12). Further consequences of this simplified concept will be discussed [58].

8.2.5. CONSERVATION OF ORBITAL SYMMETRY IN CONCERTED, STEPWISE CATALYTIC
      PROCESSES

The cyclobutanation of olefins, the formation of cyclooctatetraene from

152

four molecules of acetylene and the so-called metathesis or disproportionation
of olefins:

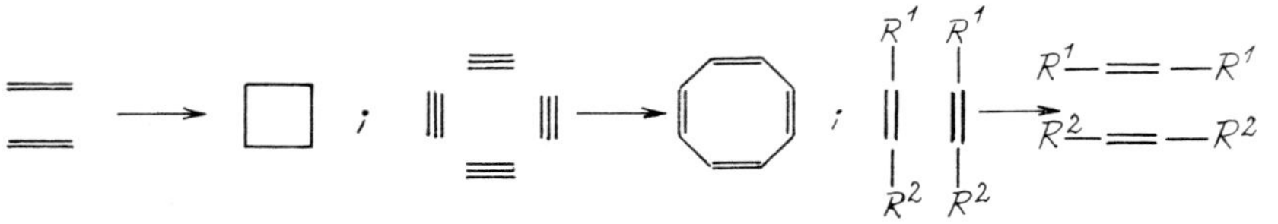

$$\equiv \longrightarrow \square \ ; \quad \begin{matrix} \equiv \\ \equiv \end{matrix} \equiv \longrightarrow \bigcirc \ ; \quad \overset{R^1 \quad R^1}{\underset{R^2 \quad R^2}{\|\quad\|}} \longrightarrow \begin{matrix} R^1 = -R^1 \\ R^2 = -R^2 \end{matrix}$$

represent overall reactions, which are forbidden processes according to the Woodward-Hoffmann rules [93] for concerted organic reactions.

These reactions, however, sometimes proceed exceptionally smoothly using
transition metals as catalysts. F. D. Mango and J. H. Schachtschneider [94,
95, 96, 97] and R. Pettit et al. [98, 99] have tried to explain these "anomalies"
with a concept called "forbidden to allowed catalysis" [100]. Their explanation
and the generalization of this concept have, however, been contradicted by several authors [47, 48, 101, 102, 103, 114].

In Wilke's research group the cycloaddition reactions, promoted by transition metals [4, 5, 9-12], were always assumed to proceed as stepwise processes.
In the cyclodi- and trimerization of butadiene to cis,cis-cycloocta-1,5-diene, cis-
1,2-divinylcyclobutane and cyclododeca-1,5,9-triene for instance, it was assumed
that the products where formed from symmetric bis-π-allyl-systems [6, 7, 81,
104, 105]; schematically:

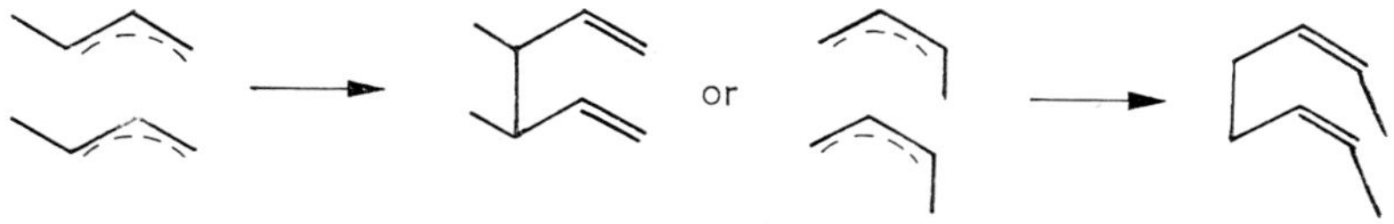

$$\text{(70)}$$

However, more recent experimental results [18, 24, 41, 48, 56] led us to the
mechanism briefly discussed in chapter 7. The first step — we believe — involves
a suprafacial C—C bond formation between a s-cis- and a s-trans 1,3-diene;
in the second step the nickel is again reduced. We count only the electrons,
which are involved in the rearrangement of the π-system see [115].

$$\text{(71)}$$

When a s-*cis*-butadiene reacts with the C_8-chain, a *cis*-π-allyl-1-hapto-allyl-C_{12}-chain is formed; schematically:

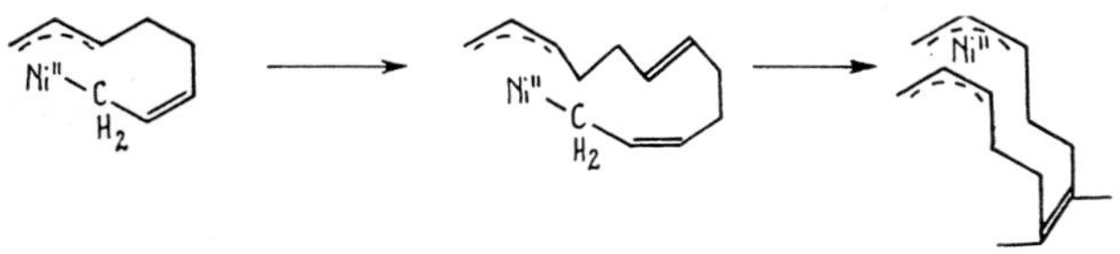

$$(72)$$

This C_{12}-chain, which is perhaps formed in an orbital symmetry controlled process, rearranges to the isolated syn,syn-C_{12}-chain.

According to F. D. Mango [100] the cyclobutanation of butadiene to *cis*-1,2-divinylcyclobutane occurs as shown in the next scheme:

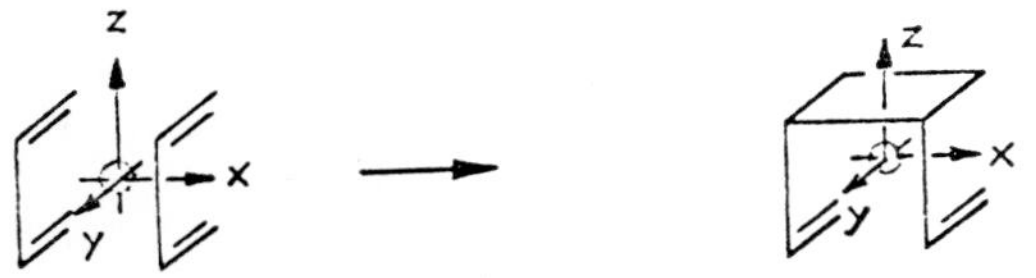

$$(73)$$

Our experiments (see chapter 6.1. Figure 17) are not in agreement with this theory, which demands that the four-membered ring VIIb should be formed from two molecules of *trans*-piperylene:

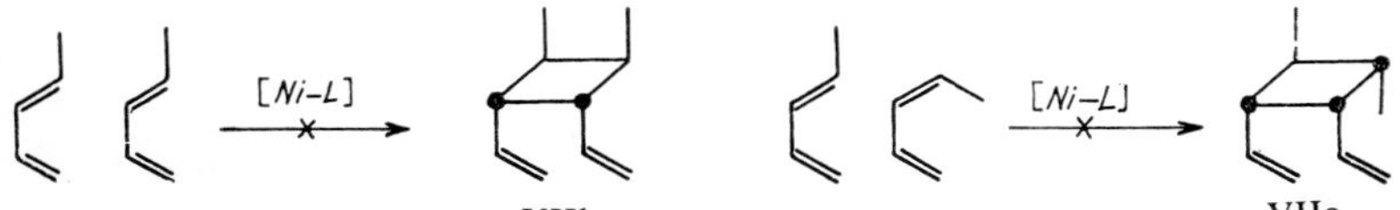

VIIb   VIIa

$$(74)$$

In fact, two molecules of *trans*-piperylene form the *cis*-1,2-divinyl-cyclobutane derivative VIIa, in which the two methyl groups are *trans* to each other. VIIb arises from one *trans*- and one *cis*-piperylene (see chapter 6.1. Figure 17) [18, 47, 48].

We assume that the Woodward-Hoffmann rules can be applied to catalytic processes, but one has to consider that e.g. the cyclobutanation of 1,3-dienes proceeds via two steps, both of which are concerted and probably orbital symmery controlled. An interesting aspect is that for instance in the case of reductive activation the lowest antibonding orbital is also occupied. In order to clarify what is meant by catalytic processes proceeding as stepwise, concerted processes, the following thermal cyclobutanation, published by Huisgen and coworkers [106], will be used as an example:

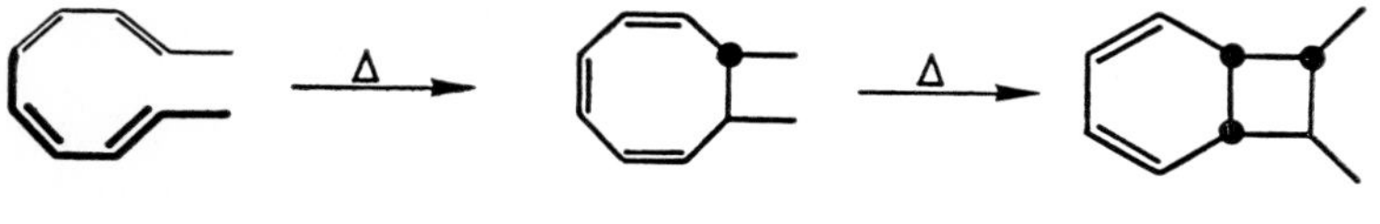

$$(75)$$

154

This stepwise, concerted cyclobutanation involves conformational changes between the first and the second step. The same may occur in stepwise, catalytic processes. Examples are the necessary rearrangements in the C_{12}-chain, before forming the cyclododeca-1,5,9-trienes (see chapter 8.2.2. equation 42).

Finally, a further catalytic reaction, which may proceed as a stepwise process, will be mentioned. The transition-state in the disproportionation of olefins is variously referred to as "quasicyclobutane" [107], "pseudocyclobutane" [108], "absorbed cyclobutane" [109], "cyclobutane" [92, 110] and "tetracarbene" [99]. As we have shown in nickelorganic chemistry, the different reactivity of π-systems within the complex depends on the oxidation-state of the metal discussed (in [58]. By assuming a reduced and oxidized form of a Mo- complex [116] it is perhaps possible, that an asymmetric complex is formed as an intermediate. In this case the principle of microscopic reversibility is not violated.

(76)

Acknowledgments

I am indebted to G. Wilke, H. Lehmkuhl and coworkers for helpful discussions and C.A. Tolman and R.S. Nyholm for communication of results prior to publication.

9. REFERENCES

[1]   H. BREIL, P. HEIMBACH, M. KRÖNER, H. MÜLLER and G. WILKE, "Makromolekulare Chem.", *69*, 18, (1963).

[2]   H. W. B. REED, "J. Chem. Soc.", 1931, (1954).

[3]   W. REPPE and W. J. SCHWECKENDIEK, "Annalen", *560*, 104, (1948).

[4]   G. WILKE et al., "Angew. Chem." *75*, 10 (1963). "Angew. Chem. Int. Ed.", *2*, 1, (1963).

[5]   G. WILKE, *Paper presented at the Xth International Conference on Coordination Chemistry*, Tokyo and Nikko, Japan (1967).

[6]   P. HEIMBACH, *Mod. Chem. Ind., Pure and Applied Chemistry Symposium, Eastbourne*, p. 223 (1968).

[7]   P. HEIMBACH, P. W. JOLLY and G. WILKE, "Adv. in Organometall. Chem." *8*, 29, (1970) Ed. F. G. A. Stone and R. West, Academic Press (New York).

[8] G. P. Chiusoli, "Aspects Homogen. Cat." 1, 77, (1970) Ed. R. Ugo, Manfredi (Milan).

[9] B. Bogdanovič, P. Heimbach, M. Kröner and G. Wilke, E. G. Hoffmann and J. Brandt, "Annalen", 727, 143, (1969).

[10] W. Brenner, P. Heimbach, H. Hey, E. W. Müller and G. Wilke, "Annalen", 727, 161, (1969).

[11] P. Heimbach and G. Wilke, "Annalen" 727, 183, (1969).

[12] W. Brenner, P. Heimbach and G. Wilke, "Annalen", 727, 194, (1969).

[13] W. Brenner, P. Heimbach, K. J. Ploner and F. Thömel, "Angew. Chem.", 81, 744, (1969); "Angew. Chem. Int. Ed.", 8, 753, (1969).

[14] P. Heimbach, "Chimia", 23, 274, (1969).

[15] A. C. Cope and E. M. Hardy, "J. Amer. Chem. Soc.", 62, 441, (1940).

[16] W. v. E. Doering and W. R. Roth, "Angew. Chem.", 75, 27, (1963); "Angew. Chem. Int. Ed.", 2, 115, (1963).

[17] E. Vogel, "Annalen", 615, 1, (1958).

[18] H. Hey, Dissertation, Ruhr-Universität Bochum (1969).

[19] G. S. Hammond and C. D. De Boer, "J. Amer. Chem. Soc.", 86, 899, (1964).

[20] P. Heimbach and W. Brenner, "Angew. Chem.", 79, 814, (1967); "Angew. Chem. Int. Ed.", 6, 800, (1967).

[21] P. Heimbach and H. Hey, "Angew. Chem.", 82, 550, (1970); "Angew. Chem. Int. Ed.", 9, 528, (1970).

[22] C. A. Grob, H. Link and P. W. Schiess, "Helv. Chim. Acta", 46, 483, (1963).

[23] P. Heimbach, "Angew. Chem.", 76, 859 (1964); "Angew. Chem. Int. Ed.", 3, 702, (1964).

[24] H. Buchholz, Dissertation, Ruhr-Universität Bochum (1971).

[25] N. Balcioglu, Dissertation, Ruhr-Universität Bochum (1970).

[26] D. W. Johnson, Dissertation, University of Wisconsin (1970).

[27] D. Craig, "J. Amer. Chem. Soc.", 72, 1678 (1950); 83, 2885, (1961).

[28] W. v. E. Doering, R. G. Buttery, R. G. Laughlin and N. Chaudhury, "J. Amer. Chem. Soc.", 78, 3224, (1956).

[29] C. Simmons, B. Richardson and I. Dvoretzky, Gas Chromatography, p. 211, (1960) Butterworths (London).

[30] T. Dvoretzky, D. B. Richardson and C. R. Durett, "Anal. Chem.", 35, 545, (1963).

[31] G. Schomburg, in preparation.

[32] G. Schomburg "Anal. Chim. Acta", 38, 45, (1967).

[33] A. Matukuma, "J. Chem. Soc.", 770, (1963).

[34] A. Matukuma, Gas Chromatography, p. 55, (1968), The Institute of Petroleum (London).

[35] G. Schomburg, "J. Chromatog.", 23, 18, (1966); "Adv. in Chromatography", 6, 211, (1968).

156

[36] K. R. KOPECKY, G. S. HAMMOND and P. A. LEERMAKERS, "J. Amer. Chem. Soc.", *84*, 1015, (1962).

[37] G. SCHOMBURG, unpublished.

[38] K. G. UNTCH and D. J. MARTIN, "J. Amer. Chem. Soc.", *87*, 3518, (1965).

[39] H. BUCHHOLZ, P. HEIMBACH, G. SCHOMBURG and G. WILKE, in preparation.

[40] E. KOVATS, "Helv. Chim. Acta", *41*, 1915, (1958); "Anal. Chem.", *36*, 31A, (1964).

[41] H. SELBECK, *Dissertation*, Ruhr-Universität Bochum (1972).

[42] P. HEIMBACH, G. SCHOMBURG, H. SELBECK and G. WILKE, unpublished.

[43] D. HENNEBERG and G. SCHOMBURG, "Z. Analyt. Chem.", *215*, 424, (1965).

[44] P. HEIMBACH, "Angew. Chem.", *22*, 967, (1968); "Angew. Chem. Int. Ed.", *7*, 882, (1968).

[45] W. FLECK, *Dissertation*, Ruhr-Universität Bochum (1971); W. FLECK, P. HEIMBACH, D. HENNEBERG and G. SCHOMBURG, in preparation.

[46] J. C. TREBELLAS, J. R. OLECHOWSKY and H. B. JONASSEN, "J. Organomet. Chem.", *6*, 412, (1966).

[47] P. HEIMBACH, W. BRENNER, H. HEY and G. WILKE, *Progress in Coordination Chemistry, Proceeding of the XI I.C.C.C., Haifa, Jerusalem*, p. 10 (1968); *Koordinationschemie der Übergangselemente Vortragsberichte Sektion C, Jena*, p. 62, (1969).

[48] P. HEIMBACH and H. HEY, "Annalen", in preparation.

[49] P. HEIMBACH, H. SELBECK and E. TROXLER, "Angew Chem.", 83, 731 (1971); "Angew. Chem. Int. Ed.", *10*, 659, (1971).

[50] P. BINGER, H. BUCHHOLZ and P. HEIMBACH, in preparation.

[51] H. J. KAMINSKY, *Dissertation*, Technische Hochschule Aachen (1962).

[52] G. WILKE and H. MÜLLER, unpublished results.

[53] P. HEIMBACH, *paper presented at Leeds-Sheffield Symposium, Leeds* (1971).

[54] P. W. JOLLY, I. TKATCHENKO and G. WILKE, "Angew. Chem.", *83*, 328, (1971); "Angew. Chem. Int. Ed.", *10*, (1971).

[55] P. W. JOLLY, I. TKATCHENKO and G. WILKE, "Angew. Chem.", *83*, 329, (1971); "Angew. Chem. Int. Ed.", *10*, (1971).

[56] H. HEY, "Angew. Chem.", *83*, 144 (1971); "Angew. Chem. Int. Ed.", *10*, (1971).

[57] F. A. COTTON, "J. Amer. Chem. Soc.", *90*, 6230, (1968).

[58] P. HEIMBACH and H. LEHMKUHL, in preparation.

[59] C. A. TOLMAN, "J. Amer. Chem. Soc.", *92*, 2953, (1970).

[60] C. A. TOLMAN, "J. Amer. Chem. Soc.", *92*, 2956, (1970).

[61] H. BÖNNEMANN, unpublished results.

[62] P. HEIMBACH, *paper presented at Leeds-Sheffield Symposium, Sheffield* (1969) and *Inorganic Symposium, Canberra* (1969).

[63] R. HUISGEN, "Angew. Chem.", *19*, *783*, (1970); "Angew. Chem. Int. Ed.", *9*, 314, (1970).

[64] F. THÖMEL, *Dissertation*, Ruhr-Universität Bochum (1970).

157

[65] P. HEIMBACH, K. J. PLONER and F. THÖMEL, "Angew. Chem.", *83*, 285, (1971); "Angew. Chem. Int. Ed.", *10*, (1971).

[66] W. RING, *Chemische Werke Hüls, Offenlegungsschrift*, DAS 1902, 274, (1970).

[67] P. HEIMBACH, W. RING, F. THÖMEL, and W. ZADV unpublished.

[68] W. REPPE, N. V. KUTEPOW and A. MAGIN, "Angew. Chem.", *81*, 717, (1969); "Angew. Chem. Int. Ed.", *8*, 727, (1969).

[69] P. HEIMBACH, unpublished results.

[70] E. KOERNER von GUSTORF, personal communication.

[71] H. BÖNNEMANN, *paper presented at Leeds-Sheffield Symposium, Sheffield* (1969).

[72] B. BOGDANOVIČ and G. WILKE, "Brennstoffchemie", *49*, 323, (1968).

[73] B. BOGDANOVIČ, B. HENC, H. G. KARMANN, H. G. NÜSSEL, D. WALTER and G. WILKE, "Industrial and Engineering Chemistry", *62*, 34, (1970).

[74] H. BÖNNEMANN, H. SCHENKLUHN and G. WILKE, unpublished results; H. BÖNNE-MANN, *paper presented at Leeds-Sheffield Symposium, Leeds* (1971).

[75] H. SCHENKLUHN, *Dissertation*, Ruhr-Universität Bochum (1971).

[76] G. HERRMANN and G. WILKE, "Angew. Chem.", *74*, 693, (1962); "Angew. Chem. Int. Ed.", *1*, 549, (1962).

[77] K. JONAS, *Dissertation*, Ruhr-Universität Bochum (1968).

[78] G. WILKE and K. JONAS, unpublished results.

[79] C. DELLIEHAUSEN, *Dissertation*, Ruhr-Universität Bochum (1968).

[80] P. HEIMBACH and G. WILKE, unpublished results.

[81] P. HEIMBACH and R. TRAUNMÜLLER, *Chemie der Metall-Olefin Komplexe*, (1970), Verlag Chemie (Weinheim).

[82] K. J. PLONER, *Dissertation*, Ruhr-Universität Bochum (1969).

[83] J. HALPERN, "Acc. Chem. Res.", *3*, 386, (1970).

[84] J. P. COLLMAN, "Acc. Chem. Res.", *1*, 136, (1968).

[85] L. VASKA and J. W. DiLUZIO, "J. Amer. Chem. Soc.", *83*, 2784, (1961).

[86] L. VASKA, "Acc. Chem. Res.", *1*, 186, (1968).

[87] K. JONAS and G. WILKE, "Angew. Chem.", *81*, 534, (1969); "Angew. Chem. Int. Ed.", *1*, 519, (1969).

[88] G. WILKE and K. JONAS, unpublished results.

[89] T. H. COFFIELD, J. KOZIKOWSKI and R. N. CLOSSON, "J. Org. Chem.", *22*, 598, (1957).

[90] a) K. A. KEBLYS and A. H. FILBEY, "J. Amer. Chem. Soc.", *82*, 4204, (1960).
b) F. CALDERAZZO, "Inorg. Chem.", *4*, 293, (1965).
c) R. J. MAWBY, F. BASOLO and R. G. PEARSON, "J. Amer. Chem. Soc.", *86*, 3994, (1964).
d) F. CALDERAZZO and F. A. COTTON, "Chimica e Industria" (Milan), *46*, 1165, (1964).

[91] G. LEFÉBVRE and Y. CHAUVIN, "Aspects Homog. Catalysis", *1*, 108, (1970). Ed. R. Ugo, Manfredi (Milan).

158

[92]   G. Henrici-Olivé and S. Olivé, *Polymerisation* (*Katalyse, Kinetik, Mechanism*), (1970), Verlag Chemie (Weinheim).

[93]   R. B. Woodward and R. Hoffmann, "Angew. Chem.", *81*, 797, (1969); "Angew. Chem. Int. Ed.", *8*, 781, (1969).

[94]   F. D. Mango and J. H. Schachtschneider, "J. Amer. Chem. Soc.", *89*, 2484, (1967).

[95]   F. D. Mango and J. H. Schachtschneider, "J. Amer. Chem. Soc.", *91*, 1030, (1969).

[96]   F. D. Mango and J. H. Schachtschneider, "J. Amer. Chem. Soc.", *93*, 1123, (1971).

[97]   F. D. Mango, "Adv. in Catalysis", *20*, 291, (1970).

[98]   R. Pettit, H. Sugahara, J. Wristers and W. Merk, "Disc. Far Soc.", *47*, 71, (1969).

[99]   G. S. Lewandos and R. Pettit, "Tetrahedron Lett.", 789, (1971).

[100] F. D. Mango, "Tetrahedron Lett.", *6*, 505, (1971).

[101] L. Cassar, P. E. Eaton and J. Halpern, "J. Amer. Chem. Soc.", *92*, 515, (1970).

[102] Th. J. Katz, N. Acton and S. Cerefice, "J. Amer. Chem. Soc.", *91*, 206 and 2405, (1969).

[103] W. Th. A. M. van der Lugt, "Tetrahedron Lett.", 2281, (1970).

[104] R. Traunmüller, *Dissertation*, Universität Wien (1969).

[105] R. Traunmüller, O. E. Polansky, P. Heimbach and G. Wilke, "Chemical Physics Letters", *3*, 300, (1969).

[106] R. Huisgen, A. Dahmen and H. Huber, "J. Amer. Chem. Soc.", *89*, 7130, (1967).

[107] C. P. C. Bradshaw, E. J. Howmann and L. Turner, "J. Catalysis", *7*, 269, (1967).

[108] E. A. Zuech, "Chem. Comm.", 1182, (1968).

[109] C. T. Adams and S. G. Brandenberger, "J. Catalysis", *13*, 360, (1969).

[110] N. Calderon, H. Y. Chen and K. W. Scott, "Tetrahedron Lett.", 3327, (1967).

[111] C.A. Tolman, Chem. Soc. Reviewes in press.

[112] H. Bönnemann, Ch. Gyrard, W. Kopp and G. Wilke, *Special Lectures XXIII. Internat. Congr. Pune Applied Chem.*, Boston, *6*, 265, (1971) Butterworths.

[113] J. Lewis and R. S. Nyholm, *Special Lectures XXIII. Internat. Congr. Pure Applied Chem.* Boston *6*, 61, (1971) Butterworths.

[114] L. Cassar, P. E. Eaton and J. Halpern, "J. Amer. Chem. Soc." *92*, 3515, (1970).

[115] H. Buchholz, P. Heimbach, H. Hey, H. Selbeck and W. Wiese, "Coord. Chem. Rev., in press.

[116] J. P. Collman, P. Eamhann and G. Dolcetti, "J. Amer. Chem. Soc.", *93*, 1788, (1971).

Chapter 4

Dimerization of Acrylic Compounds

MASANOBU HIDAI and AKIRA MISONO

Department of Industrial Chemistry, The University of Tokyo, Hongo, Tokyo, Japan

1. INTRODUCTION

The chemistry of oligomerization reactions is currently receiving considerable attention from the industrial point of view. Dimerization and co-dimerization of olefins, dienes, alkynes, and acrylic monomers which can be produced in large

quantities and at low cost by the petrochemical industry afford valuable intermediates. In particular the dimerization of acrylic compounds opens a new
field for the synthesis of bifunctional materials. There is, for example, the reductive dimerization of acrylonitrile to adiponitrile which is an important process
since the latter compound is a raw material for 6,6-nylon. Several methods for
the dimerization reaction have been discovered, but the dimerization catalyzed
by transition metal complexes has attracted much interest since the reaction
proceeds rather selectively under mild conditions.

Dimerization of acrylic compounds has been less extensively studied than
that of olefins, dienes, and alkynes. Therefore, few reviews [1] have been published
on this subject until now. The aim of this review is to give a picture of dimerization of acrylic compounds up to 1972. The reactions catalyzed by transition
metal complexes will be described in detail.

2. THERMAL AND PHOTO DIMERIZATION

Various carbonylenic compounds such as acrolein and methyl vinyl ketone
undergo thermal dimerization to give up to 55% yields of dihydropyran derivatives of the type represented by structure I [2]. Coyner and Hillman [3] carried

$$R=H \text{ or } CH_3$$

out this type of reaction with acrylonitrile at temperatures of 200-300° in the
presence of hydroquinone as a polymerization inhibitor. The only dimeric
products isolated were not heterocyclic in nature, but were found to be *cis*- and
trans-1,2-dicyanocyclobutane. The yield of acrylonitrile dimer was seriously
contaminated with tar. They proposed that this reaction proceeds through a
radical mechanism as shown below:

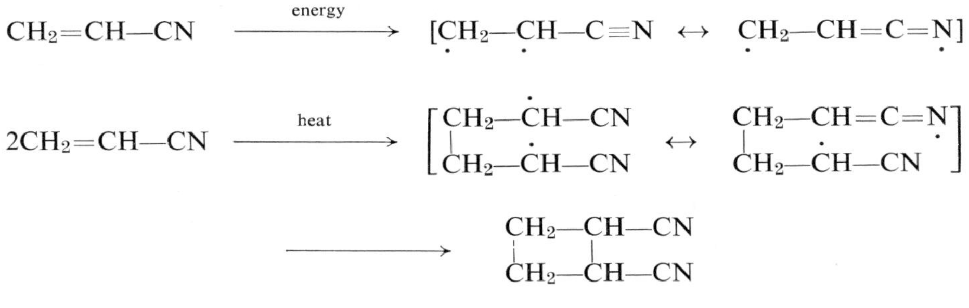

$$CH_2{=}CH{-}CN \xrightarrow{\text{energy}} [\overset{\cdot}{C}H_2{-}\overset{\cdot}{C}H{-}C{\equiv}N \leftrightarrow \overset{\cdot}{C}H_2{-}CH{=}C{=}\overset{\cdot}{N}]$$

$$2CH_2{=}CH{-}CN \xrightarrow{\text{heat}} \begin{bmatrix} CH_2{-}\overset{\cdot}{C}H{-}CN \\ CH_2{-}\overset{\cdot}{C}H{-}CN \end{bmatrix} \leftrightarrow \begin{bmatrix} CH_2{-}CH{=}C{=}\overset{\cdot}{N} \\ CH_2{-}\overset{\cdot}{C}H{-}CN \end{bmatrix}$$

$$\xrightarrow{} \begin{matrix} CH_2{-}CH{-}CN \\ | \qquad | \\ CH_2{-}CH{-}CN \end{matrix}$$

Many patents have been proposed which claimed improved processes for dimerizing acrylonitrile at faster rates and in superior conversions and yields. In one of these gaseous acrylonitrile is heated at 275-300° and pressure $\geq$ 1000 psi, either out of contact with non noble metals or in the presence of H_2O with pH < 8 to prevent a cyanoethylation reaction. The desired pH can be provided by CO_2 or any amine acceptor. A stainless steel shaker tube was charged with 120 g of acrylonitrile and 15 g of aqueous 0.001 N H_2SO_4 and the mixture was heated for 30 minutes at 275° and 1200 psi. The product was distilled to yield 88 g of acrylonitrile and 28.4 g of 1,2-dicyanocyclobutane [4]. Some patents claimed that the yield of cyclic dimer was increased by addition of CO, NO, $Ni(CO)_4$ [5a], sulfur containing compounds such as H_2S, NiS, dodecylmercaptan, and SO_2 [5b], p-tert-butylpyrocatecol [5c], or phthalic acid dibutyl ester [5d].

Photodimerization of acrylonitrile has been recently studied in order to lower the reaction temperature and prevent the polymerization. Three research groups [6] have found independently that acrylonitrile dimerizes into 1,2-dicyanocyclobutane photochemically in the presence of sensitizers such as acetophenone or benzophenone. In a typical example, a solution containing 32.2 g of acrylonitrile, 31.5 g of acetonitrile, 1 g of benzophenone and 1 g of aqueous ammonia was irradiated by a 270 W high pressure mercury lamp at 20° for 6 hr. 9.58% of the acrylonitrile was converted into cis- and trans-1,2-dicyanocyclobutane and no polymer was formed. Effective sensitizers must have their triplet excitation energy more than ca. 62 kcal/mole. Hosaka and Wakamatsu [6b] suggested the following reaction mechanism.

$$S_{(sensitizer)} \longrightarrow S^{*(1)} \longrightarrow S^{*(3)}$$

$$S^{*(3)} + CH_2=CHCN \longrightarrow S + CH_2=CHCN^{*(3)}$$

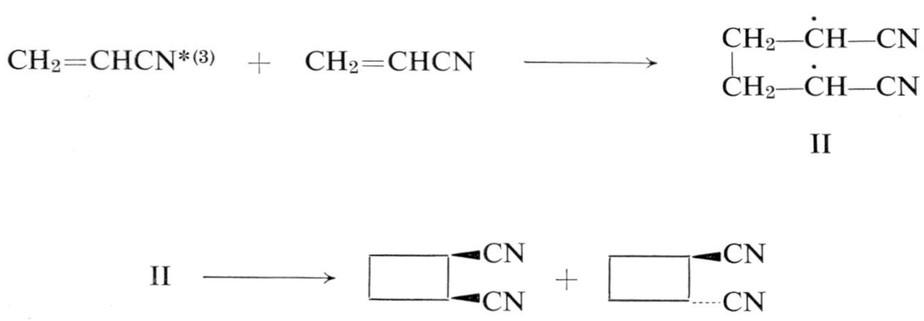

$$CH_2=CHCN^{*(3)} + CH_2=CHCN \longrightarrow \begin{array}{c} CH_2\text{---}\overset{\cdot}{C}H\text{---}CN \\ | \\ CH_2\text{---}\overset{\cdot}{C}H\text{---}CN \end{array}$$

II

$$II \longrightarrow \boxed{}\!\!\!\begin{array}{l}\text{---}CN \\ \text{---}CN\end{array} + \boxed{}\!\!\!\begin{array}{l}\text{---}CN \\ \cdots CN\end{array}$$

Probably, intramolecular coupling of the biradical II proceeds so rapidly that dimerization predominates over polymerization.

3. DIMERIZATION BY ALKALI METAL AMALGAMS
AND DIRECT ELECTROLYSIS

It has long been recognized that compounds which contain unsaturated carbon-carbon bonds in conjugation with electron acceptor groups may, under certain reducing conditions, yield a dimerized product in addition to the simple saturated monomer and some miscellaneous by-products.

$$R—CH=CH—X \xrightarrow{\text{Reduction}}
\begin{cases}
R—CH_2—CH_2—X \text{ (reduced monomer)} \\
\begin{array}{l} R—CH—CH_2—X \\ \quad\ \ | \\ R—CH—CH_2—X \end{array} \text{ (hydrodimer)}
\end{cases}$$

Here R can be H, Ph, CH_3, etc., and X can be CN, COOR, CHO, COR, COOH, $CONH_2$, etc. The chemical reducing agents that have been most often used are alkali metal amalgams or finely dispersed alkali metals in water [7], magnesium in methanol [8], and such other combinations that are strong enough to liberate hydrogen from aqueous or alcoholic solutions. Yields of the dimer product were discouragingly low, in spite of the numerous attempts prompted by the obvious commercial importance of the acrylonitrile to adiponitrile conversion. However, new impetus was provided by the publications of Knunyants and his co-workers [9] in which they claimed that acrylonitrile and other derivatives of α,β-unsaturated acids, dissolved or suspended in strong mineral acid and treated with alkali metal amalgams, which were generated electrochemically, gave good yields of "hydrodimer". Under special conditions, acrylonitrile

$$\underset{|}{\overset{|}{C}}=\underset{|}{\overset{|}{C}}—X \xrightarrow{\text{Na—Hg, K—Hg}} X—\underset{H}{\overset{|}{C}}—\overset{|}{\underset{|}{C}}—\overset{|}{\underset{|}{C}}—\underset{H}{\overset{|}{C}}—X$$

X=CN, COOR, $CONR_2$, etc.

was said to yield adiponitrile in over 60% yield; a small quantity of propionitrile was produced as a by-product but no amines or polymers were formed. The disadvantage of the method is that a large portion of the alkali metal is wasted for it reacts with the acid to give off hydrogen, instead of reacting with the acrylonitrile.

Knunyants proposed that a free radical intermediate, cyanoethyl radical, dimerized to yield the product:

$$CH_2=CHCN + H^+ + e \longrightarrow \cdot CH_2CH_2CN$$

$$2[\cdot CH_2CH_2CN] \longrightarrow NC(CH_2)_4CN$$

$$\cdot CH_2CH_2CN + H^+ + e \longrightarrow CH_3CH_2CN$$

In 1963, Baizer [10] developed a direct electrolytic method which has attracted a lot of attention and has achieved some commercial realization. The electrolysis of concentrated solutions of acrylonitrile in aqueous tetraethylammonium p-toluenesulfonate at lead or mercury cathodes and at controlled pH yields adiponitrile in virtually quantitative yields and at current efficiencies close to 100%. If the electrolyte has acrylonitrile concentrations much below 10% or contains alkali metal cations, increasing quantities of propionitrile appear as a by-product. The following mechanism was proposed by Baizer:

$$CH_2=CHCN \xrightarrow{\ \ e\ \ } [\cdot CH_2CHCN]^-[R_4N]^+$$

$$[:CH_2CHCN]^= \xleftarrow{\ \ e\ \ } \quad [\cdot CH_2CH_2CN] \ + \ OH^-$$

$$\text{III}$$

$$[\cdot CH_2CHCN]^-[R_4N]^+ \xrightarrow{\ H_2O\ } [\cdot CH_2CH_2CN] + OH^-$$

$$[\cdot CH_2CH_2CN] \xrightarrow{\ e\ } [:CH_2CH_2CN]^-$$

$$\text{IV}$$

$$\text{III} + 2H_2O \longrightarrow CH_3CH_2CN + 2OH^-$$

$$\text{III} + CH_2=CHCN \longrightarrow [NCCHCH_2CH_2CHCN]^=$$

$$[NCCHCH_2CH_2CHCN]^= + 2H_2O \longrightarrow NC(CH_2)_4CN + 2OH^-$$

$$\text{IV} + H_2O \longrightarrow CH_3CH_2CN$$

$$\text{IV} + CH_2=CHCN \longrightarrow [NCCHCH_2CH_2CH_2CN]^-$$

$$[NCCHCH_2CH_2CH_2CN]^- + H_2O \longrightarrow NC(CH_2)_4CN + OH^-$$

At the cathode acrylonitrile undergoes an over-all two electron uptake. The carbanions (III, IV) may attack either water or the β-position of highly polarized molecules of acrylonitrile which are attached to the cathode. Attack upon acrylonitrile is favoured by maintaining it in high concentrations. Interaction of the adiponitrile anion with water terminates the reaction and yields adiponitrile and OH⁻.

Electrolytic reductive dimerization of a variety of derivatives of α,β-unsaturated acids was reported, e. g., methacrylonitrile, cinnamonitrile, ethyl acrylate, ethyl maleate, di-2-ethylhexyl fumarate, N,N'-diethylacrylamide, and acrylamide [11]. Some of the results are given in Table 1.

$$\underset{\underset{R'}{|}}{\overset{\overset{R\ \ \ \ R''}{|\ \ \ \ \ |}}{C}}=C-X \quad \xrightarrow[2H_2O]{2e} \quad X-\underset{\underset{H}{|}}{\overset{\overset{R''}{|}}{C}}-\underset{\underset{R'}{|}}{\overset{\overset{R}{|}}{C}}-\underset{\underset{R'}{|}}{\overset{\overset{R}{|}}{C}}-\underset{\underset{H}{|}}{\overset{\overset{R''}{|}}{C}}-X$$

X = CN, COOR, CONR₂, etc.

This electrochemical process has been broadened to include reductive coupling of different species. The monomers are linked through their respective β-positions and there is thus provided a new synthesis of cyano esters, ester amides, etc. [12]:

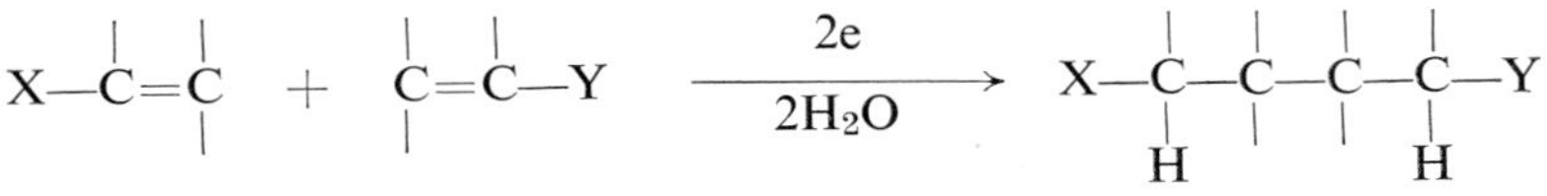

$$X-\overset{|}{C}=\overset{|}{C} \;+\; \overset{|}{C}=\overset{|}{C}-Y \xrightarrow[\text{2H}_2\text{O}]{\text{2e}} X-\overset{|}{\underset{H}{C}}-\overset{|}{C}-\overset{|}{C}-\overset{|}{\underset{H}{C}}-Y$$

X, Y = CN, COOR, CONR$_2$, etc.

Table 1

ELECTROLYTIC HYDRODIMERIZATION OF DERIVATIVES OF α, β-UNSATURATED ACIDS

Monomer	Cathode Voltage vs. S.C.E.	% Hydrodimer [1]
CH$_2$=CHCN	1.81–1.91	75–100
CH$_3$CH=CHCN	2.80–2.11	87
CN-CH=CHCN	1.42–1.60	60.3
CH$_2$=CHCOOEt	1.85	74–87
(CH$_3$)$_2$C=CHCOOEt	2.10–2.18	66.2
CH$_2$=CHCONH$_2$	1.82–2.00	39.6
CH$_2$=CHCONEt$_2$	1.91–1.95	73.3
cis $\begin{Vmatrix}\text{CHCOOEt}\\\text{CHCOOEt}\end{Vmatrix}$	1.32–1.40	61.5

[1] Based on current passed.

Arad and co-workers [13] studied the reduction of acrylonitrile by electrolysis and by reaction with alkali amalgam in acid and in neutral media. Electrolysis experiments in the presence of M$^+$ or (CH$_3$)$_4$N$^+$ in acid media yielded only hydrogen and propionitrile without any trace of adiponitrile. The amalgam experiments, on the other hand, yielded hydrogen, propionitrile, and also adiponitrile. The adiponitrile yield increased with the concentration of HCl. This was explained by assuming the protonation of acrylonitrile by the acid followed by the interaction with the alkali metal and dimerization of the free radicals formed:

$$CH_2{=}CHCN \ + \ H^+ \ \longrightarrow \ CH_2{=}CH{-}C{\equiv}NH^+$$

$$CH_2{=}CH{-}C{\equiv}NH^+ + \ M \ \longrightarrow \ [CH_2{=}CH{-}\overset{\cdot}{C}{=}NH] \ + \ M^+$$

$$\text{or} \ \ \ \ \ [{\cdot}CH_2{-}CH_2{-}CN] \ + \ M^+$$

On the other hand, in neutral media the amalgam and electrolytic reduction gave the same results. Therefore the two cases were assumed to be similar and the following mechanism containing $[{:}CH_2CHCN]^=$ as the active species was proposed:

$$CH_2{=}CHCN \ \xrightarrow{\ \ 2e\ \ } \ [{:}CH_2CHCN]^=$$

$$III$$

$$III + H_2O \longrightarrow [CH_3CHCN]^- + OH^- \xrightarrow{+ H_2O} CH_3CH_2CN + OH^-$$

$$III + CH_2{=}CHCN \ \longrightarrow \ [NCCHCH_2CH_2CHCN]^=$$

$$\xrightarrow{\ \ 2H_2O\ \ } \ NC(CH_2)_4CN \ + \ 2OH^-$$

In neutral media tetraalkylammonium salts were considered to have a specific solvating effect on the reactants. Thus, high adiponitrile yields could be obtained not only by electrolysis in the absence of alkali metal but also by reduction of an acrylonitrile solution of tetraethylammonium tosylate with alkali metal amalgam. Similar results were also reported by Matsuda [14] and in ICI patents [15].

Figeys [16] applied simple molecular orbital theory to the study of the metal and electrolytic hydrodimerization reaction of acrylonitrile and some related compounds. A mechanism based on the addition of one single electron in the potential determining step of the reduction was proposed. The sites of protonation of the radical anion and of nucleophilic attack on the neutral molecule were predicted by calculating the appropriate localization electronic densities.

4. PHOSPHINE-CATALYZED DIMERIZATION

Takashima and Price [17] reported in 1962 that while the polymerization of acrylonitrile in the presence of triphenylphosphine produces an amorphorus polymer, in the presence of ethanol and other alcohols, the principal products of the reaction are a high-melting, insoluble hexamer V, the addition product of alcohol to acrylonitrile, and a small amount of triphenylphosphine oxide.

$$CH_2=CHCN \quad \xrightarrow[\text{in EtOH}]{PPh_3} \quad (NCCH_2CH_2)_2C\underset{\overset{|}{CN}}{\overset{\overset{H}{|}}{-C}}=C\underset{\overset{|}{H}}{-}\overset{\overset{CN}{|}}{C}(CH_2CH_2CN)_2 \quad +$$

$$+ \quad C_2H_5OCH_2CH_2CN \quad + \quad Ph_3PO$$

The formation of V was considered to proceed through the intermediate 1,4-dicyano-2-butene since V was obtained from cyanoethylation of 1,4-dicyano-2-butene.

On the other hand, Rauhut and Currier [18] disclosed that treatment of alkyl acrylates with a catalytic amount of tributylphosphine in acetonitrile yields dimers of the general formula VI.

$$CH_2=CHCOOR \quad \xrightarrow[\text{in } CH_3CN]{PBu_3} \quad ROOC\overset{\overset{CH_2}{\|}}{C}CH_2CH_2COOR$$

$$VI$$

R = Me, Et, or dodecyl

Baizer and Anderson [19] attempted to apply the above procedure to acrylonitrile, but this led to a vigorous polymerization at the distillation stage. However, they found that addition of stabilized acrylonitrile to a solution of tributylphosphine in acetonitrile containing water or t-butyl alcohol followed by separation of the phosphorus compounds before isolation of the products yielded the dimer, 2-methyleneglutaronitrile VII, the trimer VIII, and a benzene-insoluble tar. In a typical experiment 50.0 g of stabilized acrylonitrile dissolved in

$$CH_2=\overset{\overset{\textstyle |}{}}{C}\underset{\overset{|}{CH_2CH_2CN}}{-CN}$$

VII

$$CH_2=\overset{\overset{\textstyle |}{}}{C}\underset{\overset{|}{CH_2CHCN}}{-CN}$$
$$\underset{CH_2CH_2CN}{|}$$

VIII

$$NCCH=CHCH_2CH_2CN$$

IX

20.7 g of t-butyl alcohol were added under nitrogen with stirring to a solution of 1.0 g of tributylphosphine in 35 ml of acetonitrile. The mixture was cooled intermittently to keep the temperature below 45° and after stirring for 3 hr at room temperature, 3.5 g of VII was obtained.

These findings are in apparent contradiction to the above report by Takashima and Price. Baizer and Anderson carried out the reaction of acrylonitrile in the presence of triphenylphosphine in t-butyl alcohol and obtained 2-methyleneglutaronitrile VII, cis- and trans-1,4-dicyano-1-butene IX and the acrylonitrile hexamer V. VII and IX were formed at about the same rate initially, but the concentration of the latter species fell off as the formation of hexamer V increased. No 1,4-dicyano-2-butene was found. From these results the following mechanism was proposed [19]:

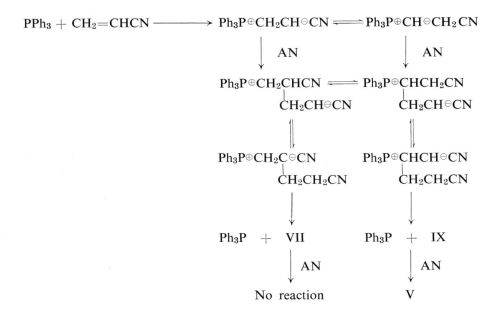

Dietsche [20] obtained similar results by using diaryl and arylalkylphosphines in *t*-butyl alcohol.

Other research groups also independently discovered the same type of dimerization reaction. Thiers and co-workers [21] described that VII is prepared by heating acrylonitrile containing 1-4 mole-% of a polymerization inhibitor in the presence of 0.4-1 mole-% of a tertiary phosphine in which the substituents are alkyl, cycloalkyl, or aryl groups. The reaction is preferably carried out in an inert atmosphere at 95-125°. For example, 16 g of acrylonitrile, 1 g of hydroquinone, and 60 ml of dioxane were refluxed with gradual addition of 0.6 g of tricyclohexylphosphine in 20 ml of dioxane for 20 min. After heating for a further 20 min 8.85 g of VII were isolated.

McClure [22] reported that the triarylphosphine-catalyzed dimerization of acrylonitrile is of significantly greater value for the synthesis of IX than implied by Baizer and Anderson [19]. Both the conversion of acrylonitrile and the yields of VII and IX vary significantly with the reaction temperature, the nature of the triarylphosphine, and the solvent. The results in Table 2 show that in *t*-butyl alcohol solvent at 175° with triphenylphosphine as catalyst, only a 45% yield of the dimers (VII and IX) at a 15% conversion of acrylonitrile is obtained. The principal by-products are an insoluble crystalline hexamer (15-20%) and a soluble polymer (25-30%). When tri(*p*-tolyl)phosphine $(CH_3C_6H_4)_3P$ is used at 160° as catalyst in place of triphenylphosphine, the acrylonitrile conversion is doubled to 30% and the yield of dimers is increased to 65%. However, the percentage of IX in the dimer decreases from 39 to 30. When triethylsilanol (Et_3SiOH) is used as the solvent in place of *t*-butyl (triphenylphosphine as catalyst at 175°) the yield of dimers is increased from 45 to 75%. Finally, with triethylsilanol as

Table 2

TRIARYLPHOSPHINE CATALYZED DIMERIZATION OF ACRYLONITRILE [22b]
(40 g of monomer)

Catalyst (1 g)	Solvent		Temp. °C	Time hrs	% conv of AN	% yield of dimers (VII + IX)	% IX in dimers
(C$_6$H$_5$)$_3$P	Me$_3$COH	(40 g)	175	8	15	45	39
(CH$_3$C$_6$H$_4$)$_3$P	Me$_3$COH	(80 g)	160	8	30	65	30
(C$_6$H$_5$)$_3$P	Et$_3$SiOH	(40 g)	175	8	16	75	40
(CH$_3$C$_6$H$_4$)$_3$P	Et$_3$SiOH	(80 g)	160	11	36	85	32

the solvent and tri(p-tolyl)-phosphine as the catalyst at 160°, an 85% yield of dimers is realized at 36% conversion of acrylonitrile. The dimer composition is 32% IX and 68% VII. Thus, by proper choice of triarylphosphine catalyst and solvent, a fourfold increase in acrylonitrile conversion over that reported by Baizer and Anderson [19] is attainable with only a slight decrease in the percentage of IX in the dimer.

McClure [22b] also discussed a mechanism of the dimerization of acrylonitrile, which was essentially identical with that proposed by Baizer and Anderson [19]. The improvement in conversion to dimers that is observed when tristolylphosphine is used (as catalyst) in place of triphenylphosphine was attributed to the slightly greater nucleophilicity of the methyl substituted phosphine. It is important that the nucleophilicity of the triarylphosphine is not too great because then only a low (< 10%) percentage of IX in the dimers is observed. Thus, tri(p-anisyl)phosphine which is significantly more nucleophilic than triphenylphosphine gives dimers containing only 9% of IX.

Support for the presence of the phosphorus ylid intermediate Ph$_3$P$^+$CH$^-$CH$_2$CN was obtained by both Oda et al. [23] and McClure [24] through the isolation of Wittig-type reaction products when the dimerization is conducted in the presence of an aromatic or aliphatic aldehyde. Thus, the reaction of acrylonitrile and triphenylphosphine with benzaldehyde at 140° afforded 23% yield of 4-phenyl-3-buten-enitrile. It was proposed that the reaction was initiated by nucleophilic addition of triphenylphosphine to a crylonitrile to form an inter-

$$Ph_3P + CH_2{=}CHCN \longrightarrow Ph_3P^{\oplus}CH_2CH^{\ominus}CN \rightleftharpoons Ph_3P^{\oplus}CH^{\ominus}CH_2CN$$

$$\xrightarrow{C_6H_5CHO} C_6H_5CH{=}CHCH_2CN$$

mediate zwitterion $Ph_3P^+CH_2CH^-CN$ which underwent a prototropic shift to form a ylid $Ph_3P^+CH^-CH_2CN$. The latter then reacted with the aldehyde in a Wittig reaction to give 4-phenyl-3-butene-nitrile.

In the phosphine-catalyzed dimerization, solvent effects are very strong. As shown in Table 2, the improvement in dimers yield that was observed when triethylsilanol was used instead of *t*-butyl alcohol is particularly noteworthy. Triethylsilanol is a stronger protolytic source than *t*-butyl alcohol. McClure [22b] showed that the solvent must not be too protolytic since the basic catalyst can then complex with the solvent and be rendered inactive. Thus, the rate of dimerization is reduced tenfold by the presence of 2,6-di-*t*-butylphenol (pKa $\cong$ 10) in the reaction mixture. It is apparent that a proper balance between nucleophilicity of the phosphine catalyst and protolytic strength of the solvent is quite critical for high yields of IX at good conversions of acrylonitrile. The combination of tri-tolylphosphine a catalyst and triethylsilanol as solvent is especially good.

The kinetics of phosphine-catalysis for the dimerization of acrylonitrile were investigated by Mueller and co-workers [25]. The selectivity and efficiency of the catalyst is a function of its basicity as well as of the stereochemical structure of the substituents on P. Thus, the least basic Ph_3P produces the highest reaction velocity, and the most basic tricyclohexylphosphine produces the lowest reaction velocity, while the stronger base Bu_3P is catalytically more active than the weaker base *iso*-Bu_3P. Optimum yields of 2-methyleneglutaronitrile are obtained with *n*-aliphatic or aromatic phosphines.

Besides the above reports, many patents have been proposed which claimed improved procedures for dimerizing acrylonitrile at faster rates and in improved conversions and yields using a variety of phosphines such as trialkylphosphine, tricycloalkylphosphine, triarylphosphine, or diphosphine as the catalyst [26]. It was recently claimed in a patent [27] that the ratio of linear to branched dimers was highly improved by the presence of a tertiary phosphine catalyst containing at least one strong electron withdrawing group and by non polar solvent dilution. Thus, acrylonitrile in *p*-dioxane containing diphenylcyanomethylenephosphine was treated for 3 hr at 150° to yield VII and IX. The conversion rate was 6.4% and the ratio of IX/VII = 18.3.

Other catalysts available for the dimerization of acrylonitrile were claimed in patents, e. g., phosphorus containing compounds with P—N bonds of the types $(RR'N)_3P$ and $(RR'N)_2PR''$ (R, R', R'' = CH_3, C_2H_5, etc.) [28] and arsenic and antimony compounds of the type R_3M (M = As, Sb) [29].

Other activated olefins besides acrylonitrile are known to dimerize in the presence of phosphine catalysts. Ethyl acrylate dimerizes in the presence of triphenylphosphine in *t*-butyl alcohol to give diethyl 2-methyleneglutarate and 1,4-dicarbethoxy-1-butene [19]. Tributylphosphine is also an effective catalyst

$$CH_2=CHCOOEt \xrightarrow{\;PPh_3\;} EtOOCCH_2CH_2\overset{\overset{\displaystyle CH_2}{\|}}{C}COOEt$$

$$+ \quad EtOOCCH_2CH_2CH=CHCOOEt$$

for the dimerization [18] [30]. Thus, 40 g of ethyl acrylate, 1.0 g of tributyl-phosphine, and 0.2 g of hydroquinone in 80 g of *t*-butyl alcohol were refluxed together at 120-130° for 7.5 hr and the mixture was distilled to give an 88% yield of diethyl 2-methyleneglutarate in addition to a trimer and low molecular weight polymer. Other phosphorus-containing compounds such as Me_2PPMe_2 [32], $Ph_2P(CH_2)n$ PPh_2 ($n = 1, 2$) [31], and $(RR'N)_3P$ (R, R' = Me, Et, etc.) [32] were also used as catalyst for the dimerization of α,β-olefinic esters such as methyl, ethyl, or butyl acrylate.

Crotononitrile is dimerized in the presence of a ditertiary phosphine such as *o*-phenylenebis(diethylphosphine) or tetramethyldiphosphine to yield prin-cipally 1,3-dicyano-2-methyl-2-pentene [33].

The catalytic reactions of methyl vinyl ketone with both trialkyl and triaryl-phosphines were studied by McClure [22*b*]. Treatment of methyl vinyl ketone with a catalytic amount of tributylphosphine in dilute triethylsilanol or *t*-butyl alcohol solution affords only a solid polymer even when carried out at 5-10°. However, when triphenylphospine is used as the catalyst at 118° in triethylsilanol solvent, a dimer, 3-methylene-2,6-heptanedione, is isolated in 78% yield with a 60% conversion of the ketone. This indicates that the reaction of the initially formed phosphonium zwitterion, $Ph_3P^+CH_2CH^-COCH_3$, with an other mole-cule of methyl vinyl ketone is very rapid and the zwitterion is not significantly transformed to a phosphonium ylid, $Ph_3P^+CH^-CH_2COCH_3$, before reacting.

$$CH_3COCH{=}CH_2 \quad \xrightarrow[\text{Et}_3\text{SiOH}]{\text{PPh}_3} \quad \underset{\overset{\|}{CH_2}}{CH_3COCCH_2CH_2COCH_3}$$

Acrylic compounds can be codimerized with fumaric or maleic esters in the presence of tertiary phosphines. For example, the reaction of 34.3 g of methyl acrylate, 67.0 g of diethyl fumalate, and 1.5 g of tricyclohexylphosphine in 100 ml of dioxane at 70° for 16 hr gave 95.7 g of 3-butene-1,2,3-tricarboxylic acid 1,2-diethyl-3-methyl ester [34].

$$CH_2{=}CHX \;+\; ROOCCH{=}CHCOOR \quad \xrightarrow{(C_6H_{11})_3P} \quad \underset{\underset{CH_2COOR}{|}}{\overset{\overset{CH_2{=}C{-}X}{|}}{CHCOOR}}$$

$$X = COOCH_3, \quad R = C_2H_5$$
$$X = COOH, \quad\;\; R = H$$
$$X = CN, \quad\quad\; R = C_4H_9$$

The reaction of equimolar amounts of acrylonitrile and ethyl acrylate with a tributylphosphine catalyst at 100° in *t*-butyl alcohol affords only one cross-condensation product, 2-carbethoxy-4-cyano-1-butene, in 48% yield (21% conversion of either reactant) despite the fact that two cross-condensates are possible. 2-Methyleneglutaronitrile (25% yield) and diethyl 2-methyleneglutarate (22% yield) are formed as by-products [22b].

5. DIMERIZATION BY TRANSITION METAL COMPLEXES

In 1940 Reppe [35] found that acetylene reacts to yield a cyclic tetramer, cyclooctatetraene, in the presence of $Ni(CN)_2$, while a cyclic trimer, benzene, is formed in the presence of $Ni(CO)_2(PPh_3)_2$. This cyclooligomerization of acetylene by Reppe catalysts may be historically the beginning of the selective oligomerization catalyzed by transition metal complexes. After this work, various kinds of transition metal catalysts were found for dimerization and trimerization of alkynes, 1,3-conjugated dienes, and olefins. For example, ethylene is dimerized by rhodium chloride in ethanol. Cramer [36] proposed the reaction cycle as shown below from a variety of interlocking physical and chemical evidence re-

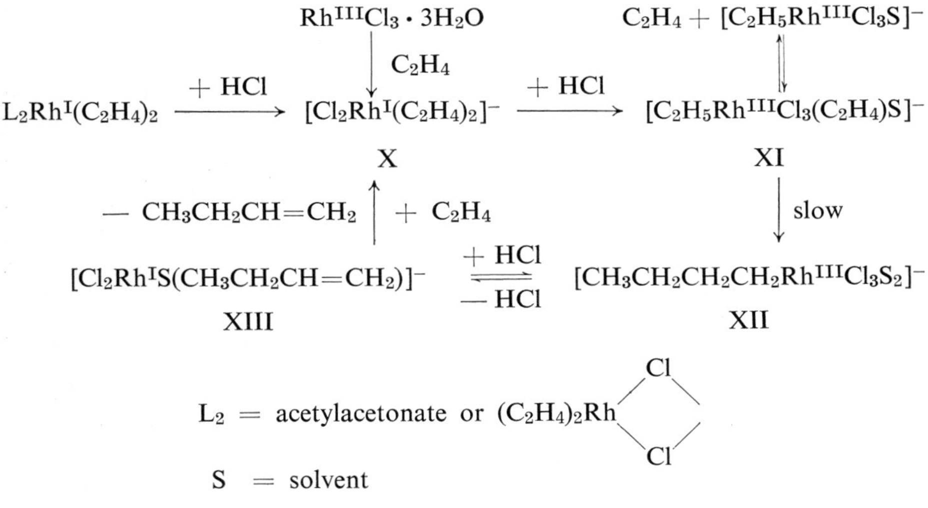

$$Rh^{III}Cl_3 \cdot 3H_2O \qquad\qquad C_2H_4 + [C_2H_5Rh^{III}Cl_3S]^-$$

$$L_2Rh^I(C_2H_4)_2 \xrightarrow{\; + HCl \;} [Cl_2Rh^I(C_2H_4)_2]^- \xrightarrow{\; + HCl \;} [C_2H_5Rh^{III}Cl_3(C_2H_4)S]^-$$

$$X \qquad\qquad XI$$

$$-CH_3CH_2CH{=}CH_2 \uparrow + C_2H_4 \qquad\qquad \text{slow}$$

$$[Cl_2Rh^I S(CH_3CH_2CH{=}CH_2)]^- \underset{-HCl}{\overset{+HCl}{\rightleftharpoons}} [CH_3CH_2CH_2CH_2Rh^{III}Cl_3S_2]^-$$

$$XIII \qquad\qquad XII$$

$$L_2 = \text{acetylacetonate or } (C_2H_4)_2Rh \overset{Cl}{\underset{Cl}{\diamond}}$$

$$S = \text{solvent}$$

lating to the intermediates and to the individual reactions. A bis ethylene complex of monovalent rhodium X is rapidly converted by oxidative addition of HCl into a σ-ethylrhodium compound XI, which rearranges by a slow, rate-determining chain growth reaction, giving an σ-butylrhodium complex XII. The complex XII decomposes rapidly through loss of HCl to give a 1-butene complex of monovalent rhodium XIII. Coordinated 1-butene and solvent in XIII are rapidly displaced by ethylene reforming the initial rhodium complex X.

The dimerization of activated olefins such as acrylic derivatives has been far less extensively studied than that of olefins, 1,3-conjugated dienes, and alkynes.

It is well known that the coordination of substrates to a transition metal is an indispensable step in dimerizations or trimerizations catalyzed by transition metal complexes as it is in other reactions such as hydrogenation, polymerization and hydroformylation.

Many transition metal complexes coordinated with acrylic compounds are known. For example, a series of iron carbonyl complexes of the type $Fe(CO)_4$ $(CH_2=CH—X)$ ($X = CHO$, $CONH_2$, $COOH$, $COOCH_3$, etc.) are prepared in good yields by the reaction of acrylic compounds with $Fe_2(CO)_9$ [37]. These compounds are yellow crystalline solids and acrylic compounds are considered to be bonded to the iron atom through the $C=C$ bond since the $C=C$ stretching frequencies shift to lower frequencies (1450-1500 cm^{-1}). It should be noted that acrylic compounds have a functional group in addition to the $C=C$ bond. This indicates that several ways of coordination of acrylic compounds are possible in addition to the coordination through the $C=C$ bond described above. For example, acrylonitrile has the $C=C$ bond and the $C\equiv N$ bond, and two kinds of coordination are known at present. One is the coordination through the nitrogen lone pair in the case of complexes such as $(CH_2=CHCN)_2PdCl_2$ [38a], $(CH_2=CHCN)_2M(CO)_4$ ($M = Cr$, Mo, W) [39], and $TiCl_3(CH_2=CHCN)_3$ [40]. The other is the coordination through the olefinic $C=C$ bond in the case of complexes such as $Ni(CH_2=CHCN)_2$ [38b], $Fe(CO)_4(CH_2=CHCN)$ [41] and $M(CO)_3(CH_2=CHCN)$ ($M = Mo$, W) [39]. In the former case, the nitrile stretching frequency is shifted to a value higher than that in free acrylonitrile, while the $C=C$ stretching frequency is essentially unchanged from its position in free acrylonitrile. In the latter case, the $C=C$ stretching frequency shifts to a lower frequency while the nitrile stretching frequency is essentially unchanged. These coordination differences are also clearly demonstrated by the measurements of n.m.r. spectra. The spectra of the former complexes are essentially identical with that of free acrylonitrile, except that the resonance lines are shifted slightly upfield. The spectra of the latter complexes, on the other hand, show a greatly perturbed resonance pattern upfield.

Guttenberger and Strohmeier [42] have reported that in acrylonitrile-π-arene-chromiun dicarbonyls, the manner of attachement of acrylonitrile to the metal changes from coordination through the nitrogen lone pair to that through the $C=C$ bond when the π-donor character of the chromium is increased by varying the arene.

In summary, acrylic compounds have several ways of coordination which depend upon the valency state of the metal, and electronic and steric factors of other ligands. It seems that the coordination through a $C=C$ bond may be essential for the dimerization of acrylic compounds as in the case of olefin dimerization.

5.1. Dimerization by iron and cobalt carbonyl complexes

As described above, many transition metal carbonyls easily react with acrylic compounds to yield complexes . In 1964-1965, three research groups independently discovered that systems derived from iron or cobalt carbonyl are effective for the non selective hydrodimerization of acrylonitrile.

A Rhone-Poulenc patent [43] claimed that the system composed of iron pentacarbonyl and an aqueous alkaline solution is effective for the formation of adiponitrile from acrylonitrile. For instance, 20 ml of acrylonitrile, in the presence of 2.8 ml of $Fe(CO)_5$, 3.6 g of aqueous 50 wt% NaOH solution, and 250 mg of hydroquinone under 85 atm of hydrogen, gives, after 15 hr at 100°, 1.85 g of adiponitrile and 4.7 g of polymerized residue. Adiponitrile is also prepared by heating acrylonitrile under hydrogen pressure in the presence of the complex $(CH_2=CHCN)_2Co(CO)_2$ obtained by reacting dicobaltoctacarbonyl with acrylonitrile in cyclohexane [44].

A Du Pont patent [45] claimed that acrylonitrile is hydrodimerized, i. e., dimerized and reduced, to form 2-methylglutaronitrile by mixing acrylonitrile with cobalt carbonyl hydride and water in the presence of carbon monoxide at a pressure of at least about 1000 psi and at a temperature within the range of 130-300°. The cobalt complex is prepared *in situ* from reduced cobalt, carbon monoxide, and water. Thus, 30 ml of acrylonitrile in 125 ml of acetone, in the presence of 5 g of cobalt (the reduced oxide), 18 ml of H_2O, and 0.5 g of iodine under 6000 psi of carbon monoxide, gives, after 2 hr at 175°, 4.5 g of 2-methylglutaronitrile. Methacrylic acid esters are also dimerized to 2,2,4-trimethylglutaric acid esters by using similar catalyst systems [46]. Thus, 30 ml of methyl methacrylate, 125 ml of acetone, 18 ml of H_2O, 0.6 g of ruthenium trichloride, and 0.1 g of iodine were reacted under 6000 psi of carbon monoxide. After 2 hr at 175°, 6.4 g of dimethyl 2,2,4-trimethylglutarate was obtained.

Misono and co-workers [47] found that acrylonitrile is converted at elevated temperatures to adiponitrile and 2-methylglutaronitrile in addition to propionitrile in an iron or cobalt carbonyl-alkali-water system, an iron or cobalt carbonyl-sodium borohydride system, or a cobalt carbonyl-hydrogen system. The hydrodimers were not obtained catalytically as described in above patents. Use of the system containing an aqueous alkali gave ethylene cyanolydrin and 2,2'-dicyanoethylether as by-products.

The solutions obtained by treating iron pentacarbonyl with aqueous alkali are known to contain the $[HFe(CO)_4]^-$ and/or the $[Fe(CO)_4]^{2-}$ ions, depending on the amount of alkali used, as shown in the following equations [48]:

$$Fe(CO)_5 + 3NaOH = NaHFe(CO)_4 + Na_2CO_3 + H_2O$$

$$Fe(CO)_5 + 4NaOH = Na_2Fe(CO)_4 + Na_2CO_3 + 2H_2O$$

$$Na_2Fe(CO)_4 + H_2O = NaOH + NaHFe(CO)_4$$

Effects of the molar ratio of NaOH to $Fe(CO)_5$ on the hydrodimerization of acrylonitrile were studied by Misono and co-workers [47]. The yield (based on the acrylonitrile initially fed) of the hydrodimers reached a maximum at a ratio of $NaOH/Fe(CO)_5 = 3$. When the molar ratio became more than 3, the yield decreased. They considered that a hydride complex such as $[HFe(CO)_4]^-$

plays an important role in the formation of hydrodimers. Support for this view was obtained by the reactions of acrylonitrile with isolated salts of $[HFe(CO)_4]^-$, $[HFe_2(CO)_9]^-$, and $[HFe_3(CO)_{11}]^-$, giving the hydrodimers. On the other hand, the reaction of acrylonitrile with hydridobis(dimethyl- or diphenylglyoximato)pyridinecobalt(III) at —20 or —40° gave 1-cyanoethyl complexes which reacted with acrylonitrile at 110° under hydrogen pressure to give 2-methylglutaronitrile [49]. Treatment of ethyl acrylate with a hydrido complex $[NEt_3H]$ $[HFe_3(CO)_{11}]$ in the presence of 3-chloropropionitrile gave ethyl 5-cyanovalerate in addition to ethyl adipate. Based on these results, the following reaction scheme was proposed for the hydrodimerization of acrylonitrile, where L_n represents the ligands other than hydrogen [47]:

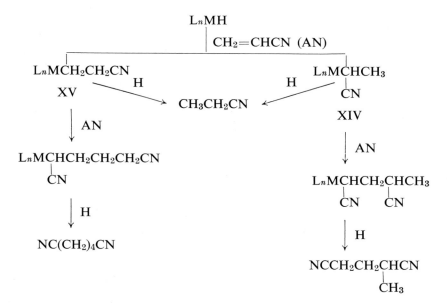

In the first step, the 1- or 2-cyanoethyl complex (XIV or XV) is formed by the addition of acrylonitrile to a hydrido complex. The 2-cyanoethyl complex is also formed by the reaction of 3-chloropropionitrile with a hydrido complex. The direction of the addition of acrylonitrile to a hydrido complex in this step determines whether the structure of the hydrodimer is branched or linear. But factors influencing the 1-cyanoethyl/2-cyanoethyl ratio are, up to now, not fully understood.

Thiers and his co-workers [50] claimed in a patent that cyanoethyl complexes of the type $MFe(CO)_3(CH_2CH_2CN)$ (M = Li, Na, K, or NH$_4$) were isolated from the reaction solution containing $Fe(CO)_5$, an alkali, water, and acrylonitrile.

Other patents [51] claimed the use of irradiated $Fe(CO)_5$ in the presence of water. For instance, 96 g of acrylonitrile in the presence of water and $Fe(CO)_5$ (40 g) were heated, after 3 hr of irradiation, at 110° for 5 hr to give 20 g of adiponitrile, 0.93 g of propionitrile, and 0.37 g of 2-methylglutaronitrile. Formation of ethylene cyanohydrin, 2,2'-dicyanoethyl ether, and polymers was remarkably depressed. A somewhat similar process has been claimed to dimerize ethyl acrylate to diethyl adipate [51b].

5.2. Dimerization by rhodium, ruthenium, and palladium complexes

Alderson and his co-workers [52] found the addition of ethylene and propylene to such dienes as butadiene, isoprene, and 1,3-pentadiene, and the dimerization of ethylene to butenes and butadiene to 2,4,6-octatriene were catalyzed by rhodium and ruthenium chlorides. The linear dimerization of an acrylic compound was also accomplished by the same catalysts. Thus, a high yield of dimethyl 2-hexenedioate was obtained by heating a solution of methyl acrylate in methanol to 140° in the presence of rhodium chloride. Essentially no higher molecular weight products were formed. The dimerization was also accomplished in 48% yield with ruthenium chloride at 120°. Surprisingly, addition of a small amount of ethylene to the latter system permitted the use of a much lower temperature, i. e., 150°, and gave a 56% yield of dimer and a low conversion to the ethylene-methyl acrylate dimer (methyl 3-pentenoate).

$$CH_2=CHCOOCH_3 \xrightarrow[\text{in } CH_3OH]{RhCl_3 \cdot 3H_2O} CH_3OOCCH=CHCH_2CH_2COOCH_3$$

Reaction of methyl acrylate with excess ethylene at 150° in the presence of ruthenium chloride gives a high yield of methyl esters of linear monounsaturated acids. The principal product is methyl 3-pentenoate (47% yield). Esters of acids containing seven and nine carbon atoms are also isolated in yields of 12 and 9%,

$$CH_2=CH_2 + CH_2=CHCOOCH_3 \longrightarrow [CH_2=CH-CH_2-CH_2-COOCH_3]$$

$$CH_2=CH_2$$

$$[CH_2=CH-(CH_2)_4-COOCH_3] \quad CH_3-CH=CH-CH_2-COOCH_3$$

etc.

respectively. Other products include the linear dimer of methyl acrylate (23% yield) and olefins formed by condensation of ethylene. Rhodium chloride as a catalyst for this reaction at 120° gives a mixture of esters similar to that obtained using ruthenium chloride [52].

Although the catalytic behavior of ruthenium is similar to that of rhodium in the presence of various olefinic compounds, the main difference between the two metals arises from their behaviour in the dimerization of acrylonitrile. Rhodium does not dimerize acrylonitrile. Treatment of ethanolic rhodium trichloride with acrylonitrile produces a voluminous yellow precipitate of approximate composition XVI in 87% yield. The complex dissolves in warm pyridine to give compound XVII in 53% yield [53].

$$[(CH_2{=}CHCN)_2RhCl_2]_n$$

XVI

$$\underset{\underset{\underset{CH_3}{|}}{CHCN}}{(py)_3RhCl_2}$$

XVII

However, ruthenium trichloride dimerizes acrylonitrile, under a hydrogen atmosphere, into *cis*- and *trans*-1,4-dicyano-1-butene and adiponitrile; the reaction is not catalytic under a nitrogen atmosphere [57] [54] [55] [56]. Nonetheless, under an ethylene atmosphere considerable amounts of dimers are obtained [55*b*]. A patent [58] claimed that under a nitrogen atmosphere and in the presence of various organic compounds a minor quantity of 1,4-dicyano-1,3-butadiene, which is the product of an oxidizing coupling, is formed.

$$CH_2{=}CHCN \xrightarrow[\text{under } H_2]{RuCl_3 \cdot 3H_2O} \begin{array}{ll} NCCH{=}CHCH_2CH_2CN & (\textit{cis-} \text{ and } \textit{trans}) \\ NC(CH_2)_4CN \\ CH_3CH_2CN \end{array}$$

The dimerization of acrylonitrile at temperatures of 100-160° under moderate hydrogen pressures gives a 50-65% yield of dimers and a 35-50% yield of propionitrile. Below 100° the reaction rate is low and above 180° the yield of dimers strongly decreases. An increase in hydrogen pressure favours the dimerization, particularly to form adiponitrile, and the formation of propionitrile, but there is a pressure beyond which the yield of dimers decreases and propionitrile becomes the main product. The optimum appears to be between 5 and 20 atm. The reaction is usually carried out in pure acrylonitrile or sometimes in the presence of alcohols which promote the reaction. Besides hydrated ruthenium trichloride $RuCl_3 \cdot 3H_2O$, other ruthenium complexes such as dichloro(dodeca-2,6,10-triene-1,12-diyl)ruthenium(IV) $RuCl_2(C_{12}H_{18})$ and ruthenium(III) acetylacetonate $Ru(acac)_3$ are also used as effective catalysts for the dimerization. However, all reactions of acrylonitrile are inhibited by the use of dichloro(dicarbonyl)bis(pyridine)ruthenium(II) $RuCl_2(CO)_2(py)_2$ in which carbon monoxide and pyridine are strongly coordinated to the ruthenium atom. The use of dichlorotetrakis(triphenylphosphine)ruthenium(II) $RuCl_2(PPh_3)_4$ and dichlorotetrakis(triphenylphosphite)ruthenium(II) $RuCl_2[P(OPh)_3]_4$ gives propionitrile as the main product [54*a*] [54*b*] [55*c*] [55*d*] [59]. Some patents claimed heterogeneous reactions in a gas or liquid phase using ruthenium chloride or powdered ruthenium metal supported on charcoal or alumina as the catalysts [57].

Although the hydrogen requirement of the dimerization suggests that ruthenium hydride complexes are involved as intermediates, the details of the mechanism of the reaction are not well understood. Three different reaction mechanisms have been proposed.

McClure and his co-workers [59] studied the catalyst system composed of $(Ph_3P)_3RuCl_2$ and N-methylpyrrolidine. The addition of the latter compound (10-20 moles per mole of ruthenium complex) produces a 2 to 3-fold increase in the conversion of acrylonitrile to propionitrile, while the yield of dimers is unaffected. Reaction of $(Ph_3P)_3RuCl_2$ with hydrogen (80 psi) in the presence of N-methylpyrrolidine (10 moles/mole of ruthenium) at 25° affords $(Ph_3P)_3RuClH$ XVIII in a good yield. The non-catalytic reaction of XVIII with acrylonitrile in benzene diluent in the presence of the amine at 25° affords propionitrile and a mixture of 2-methyleneglutaronitrile, adiponitrile, and 1,4-dicyano-1-butene. From the reaction of XVIII with 1.5-2.0 molar equivalents of acrylonitrile at 40-60° a ruthenium complex, tentatively assigned to the dimeric structure $[(Ph_3P)_2RuCl(CH_2=CHCN)]_2$ XIX, is isolated, which is reconverted to XVIII on treatment with hydrogen and amine. In the presence of the amine and hydrogen, both XVIII and XIX catalyze the reaction with acrylonitrile at 110-115° to give conversions and product compositions that are very similar to those observed with $(Ph_3P)_3RuCl_2$. Furthermore, in the absence of amine the product compositions observed with XVIII and XIX are almost identical. Based on these results, a mechanistic scheme employing compounds similar to XVIII and XIX as intermediates was formulated in a general manner for a halogen containing ruthenium catalyst as shown below [59]:

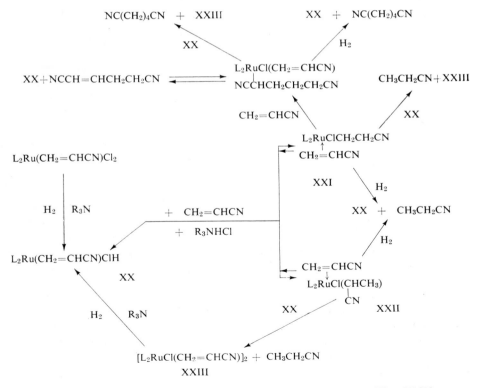

L = PPh_3 or $CH_2=CHCN$

Propionitrile formation occurs by reaction of XXI and XXII with either XX or hydrogen. When propionitrile is formed by the reaction of XXI and XXII with XX, the ruthenium complex XXIII (similar to XIX) is the co-product. Reconversion of XXIII to XX which is essential to maintain catalytic activity is apparently promoted by the presence of a tertiary amine. 1,4-Dicyano-1-butene is probably formed by an insertion of a π-bonded acrylonitrile into the 2-cyanoethyl ruthenium bond of XXII followed by elimination of XX.

Billig and his co-workers [60] reported the effect on the ruthenium chloride catalyzed dimerization of adding $SnCl_2$ (or Et_4NSnCl_3) together with bifunctional amines such as N-methylmorpholine, and/or bifunctional alcohols such as methylcellosolve. The net result of addition of these ligands is a lowering of the minimum hydrogen pressure requirement to 85 psi H_2 at 150°. In order to elucidate the role of hydrogen, they carried out a series of dimerizations under a deuterium atmosphere. Mass and NMR spectral analyses indicated that in the recovered partially deuteriated acrylonitrile deuterium was incorporated with a high degree of specificity, *trans* and β to the cyano-group, and the propionitrile formed in the reaction was largely perprotio (77.5%) and contained only 20.3% of monodeuteriopropionitrile. A mechanism which is consitent with the above data was proposed as follows (ligands are omitted for sake of clarity).

$$Ru^{II}(CH_2=CHCN) \rightleftharpoons HRu^{IV}-CH=CHCN \longrightarrow$$
$$XXIV$$

$$\xrightarrow{CH_2=CHCN} NCCH_2CH_2-Ru^{IV}-CH=CHCN \xrightarrow{CH_2=CHCN}$$

$$\xrightarrow{CH_2=CHCN} NCCH=CHCH_2CH_2CN + Ru^{II}(CH_2=CHCN)$$

$$NCCH_2CH_2-Ru^{IV}-CH=CHCN + HRu^{IV}-CH=CHCN \longrightarrow$$

$$\longrightarrow CH_3CH_2CN + NCCH=CH-Ru-Ru-CH=CHCN$$

$$NCCH=CH-Ru-Ru-CH=CHCN + H_2 \longrightarrow 2HRu^{IV}-CH=CHCN$$

A ruthenium hydride complex XXIV as an active species is formed by an oxidative addition of the vinyl-hydrogen bond of acrylonitrile. The primary role of hydrogen in the proposed scheme is for the regeneration of the active catalyst.

Misono and his co-workers [55] studied the dimerization of acrylonitrile catalyzed by hydrated ruthenium chloride in methanol or ethanol. A ruthenium acrylonitrile complex $RuCl_2(CH_2=CHCN)_3$ was obtained by heating an ethanolic solution of $RuCl_3 \cdot 3H_2O$ under reflux with a large excess of acrylonitrile under nitrogen. The infrared spectrum of the complex showed the coordination of acrylonitrile through the nitrogen lone pair (σ-bond) [55a] [55d]. Similar acrylonitrile complexes $RuCl_2(CH_2=CHCN)_4$ and $RuCl_2(CH_2=CHCN)_3(H_2O)$ were reported by McClure et al. [59] and in a Rhone-Poulenc patent [54d]. These acrylonitrile complexes catalytically dimerize acrylonitrile under a hydrogen atmosphere and can be considered as the intermediates in the dimerization. The reaction of acrylonitrile with $RuCl_2(PPh_3)_3$ gives a complex $RuCl_2(PPh_3)_2(CH_2{=}CHCN)_2$, in which acrylonitrile is attached to the metal by the C=C bond. However, the reaction in the presence of triphenylphosphite affords a σ-bonded complex $RuCl_2[P(OPh)_3]_3(CH_2{=}CHCN)$. This indicates that the coordination of acrylonitrile changes with the π donor ability of ruthenium in a similar way to the acrylonitrile-πarene chromium carbonyls described before. The dimerization reactions were also investigated from a kinetic point of view, which indicated that alcohol used as the solvent takes part in the elementary reactions. A mechanism was proposed in which the manner of attachment of acrylonitrile changes from the coordination through the nitrogen lone pair to that through the C=C bond by the coordination of the solvent (alcohol) [55d]:

$$RuCl_2(RCN)_n + H_2 \;\rightleftharpoons\; RuHCl(RCN)_n + HCl \qquad (n = 3 \text{ or } 4)$$

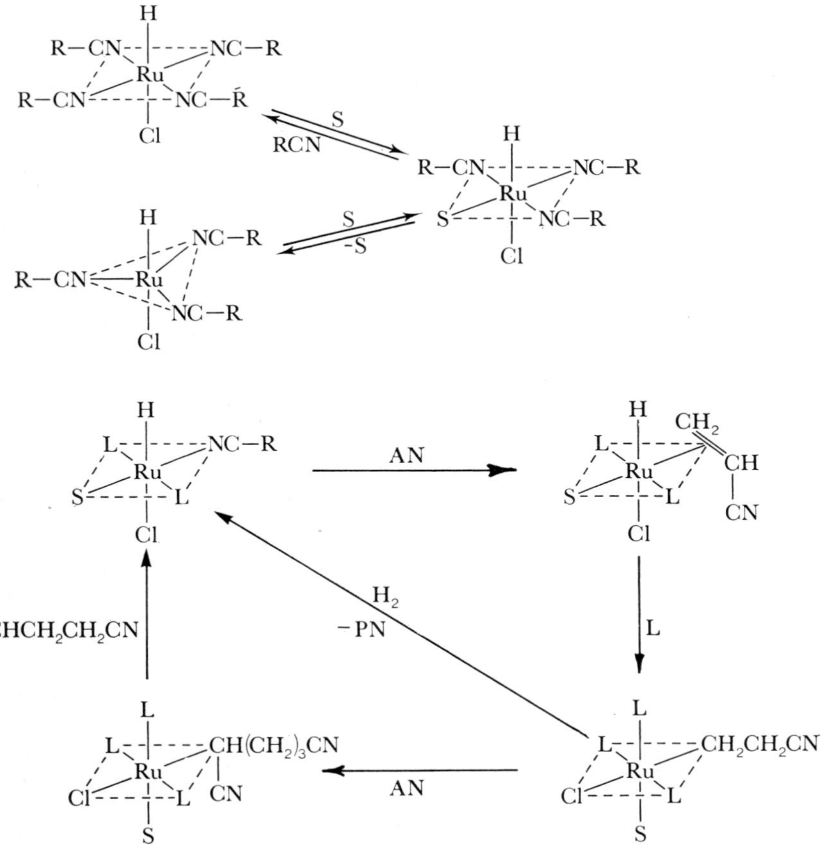

$RCN = CH_2{=}CHCN$ (AN), CH_3CH_2CN (PN),   or Dimers
$S =$ alcohol $L = RCN$

The codimerization of acrylonitrile and methyl acrylate is also catalyzed by ruthenium complexes under a hydrogen atmosphere. Besides the hydrodimerization products of each monomer, up to 50% of the head-to-head codimerization products are formed, i. e. methyl cyanovalerate $NCCH_2CH_2CH_2CH_2COOCH_3$ and methyl 5-cyano-4-pentenoate $NCCH=CHCH_2CH_2COOCH_3$ [61].

Acrylamide and its N-monosubstituted derivatives in an alcoholic solution are dimerized by rhodium trichloride into the corresponding *trans*-α-hydromuconamides, but in the case of N, N'-disubstituted derivateves, the cleavage of the amide groups takes place [62]. The reaction proceeds in heterogeneous systems.

$$2CH_2=CHCONHR \quad \xrightarrow[\text{in ROH}]{RhCl_3} \quad \underset{H}{\overset{RNHOC}{}} C=C \underset{CH_2CH_2CONHR}{\overset{H}{}}$$

The rhodium ion is reduced to form a mirror and the yield of dimer is usually 2.5-5.5 mole/per rhodium gram atom. The infrared analysis of acrylamide complexes showed that acrylamide and its N-monosubstituted derivatives are bonded to the metal through the C=C bond while N, N'-disubstituted acrylamides are coordinated to the metal at the amide group. The cleavage of the amide group instead of dimerization in the case of N, N'-disubstituted acrylamides was explained by these facts.

Dichlorobis(benzonitrile)palladium(II) $(PhCN)_2PdCl_2$ is an effective catalyst for the dimerization of methyl acrylate [63]. Thus a high yield (93%) of dimers was obtained when methyl acrylate was heated with the palladium complex at 113° for 23 hr. The product contained approximately 90% linear dimers, with dimethyl *trans*-2-hexenedioate (67%) as the maj hor isomer, while the next most prevalent isomer was dimethyl *trans*-3-hexenedioate. The complex also catalyzes the reaction of methyl acrylate and ethylene to form codimers in 47%

$$CH_2=CH_2 \quad + \quad CH_2=CHCOOCH_3 \quad \xrightarrow[105/36 \text{ hr}]{(PhCN)_2PdCl_2}$$

$$\underset{H}{\overset{CH_3}{}} C=C \underset{CH_2COOCH_3}{\overset{H}{}} \qquad \underset{H}{\overset{CH_3}{}} C=C \underset{H}{\overset{CH_2COOCH_3}{}} \qquad \underset{H}{\overset{C_2H_5}{}} C=C \underset{COOCH_3}{\overset{H}{}}$$

$$(60\%) \qquad\qquad (10\%) \qquad\qquad (30\%)$$

yield. Styrene and methyl acrylate yield a codimer in 26% yield.

$$CH_2=CHCOOCH_3 + PhCH=CH_2 \quad \xrightarrow[110/24 \text{ hr}]{(PhCN)_2PdCl_2}$$

$$\xrightarrow{\hspace{2cm}} \quad \underset{H}{\overset{Ph}{}} C=C \underset{CH_2CH_2COOCH_3}{\overset{H}{}}$$

5.3. Dimerization by other transition metal complexes

Acrylic and methacrylic acids are dimerized into 3-methylglutaconic acid and 2-methylene-4,4-dimethylglutaric acid, respectively, in the presence of pentacyanocobaltate(II) at elevated temperatures under nitrogen. The reaction under a hydrogen atmosphere affords the corresponding hydrogenated dimer as well as the corresponding hydrogenated monomer [64]:

$$CH_2=CHCOOH \xrightarrow{[Co^{II}(CN)_5]^{3-}} \begin{cases} \xrightarrow[125^\circ]{H_2} HOOCCH_2CH_2\overset{\overset{\displaystyle CH_3}{|}}{C}HCOOH \\[2mm] \qquad\qquad\qquad\qquad + CH_3CH_2COOH \\[2mm] \xrightarrow[125^\circ]{N_2} HOOCCHCH_2=\underset{\underset{\displaystyle CH_3}{|}}{C}COOH \end{cases}$$

Chiusoli and his co-workers [65] reported a novel stoichiometric procedure for dimerizing acrylonitrile to adiponitrile, based on the use of powdered manganese and cobalt chloride in dimethylformamide, according to the following two-step reaction:

$$2CH_2=CHCN + CoCl_2 + Mn \xrightarrow{DMF} (CH_2=CHCN)_2Co(DMF)_n + MnCl_2$$

$$(CH_2=CHCN)_2Co(DMF)_n + H_2S \longrightarrow nDMF + CoS + NC(CH_2)_4CN$$

The yield, based on converted acrylonitrile, exceeds 95% and a trace of 2-methylglutaronitrile is formed. The following intermediate was proposed:

$$(DMF)_2Co \begin{matrix} H & CH=CHCN \\ \diagdown & \diagup \\ | & \\ \diagup & \diagdown \\ H & CH=CHCN \end{matrix}$$

Ethyl crotonate undergoes fast and selective dimerization at 110° in the presence of a potassium-benzyl potassium catalyst to form the diethyl ester of 2-ethylidene-3-methylglutaric acid in 90% yield. Shabtai and Pines [66] proposed that the vinylic hydrogen at the α-position in respect to the ester group of ethyl crotonate is more acidic than the allylic methyl hydrogens and that dimerization is initiated by metalation at the former position.

$$C_2H_5OOCCH=CHCH_3 \underset{RH}{\overset{R^-}{\rightleftharpoons}} C_2H_5OOCC^\ominus=CHCH_3$$

$$CH_3CH=C^\ominus COOC_2H_5 + CH_3CH=CHCOOC_2H_5 \rightleftharpoons$$

$$C_2H_5OOC\overset{\overset{\displaystyle CH_3}{|}}{\underset{\underset{\displaystyle CH_3}{|}}{\overset{\displaystyle CH}{\underset{\displaystyle \,}{C}}}}CHCH^\ominus COOC_2H_5 \underset{R^\ominus}{\overset{RH}{\rightleftharpoons}} C_2H_5OOC\overset{\overset{\displaystyle CH_3}{|}}{\underset{\underset{\displaystyle CH_3}{|}}{\overset{\displaystyle CH}{\underset{\displaystyle \,}{C}}}}CHCH_2COOC_2H_5$$

β-Alkyl α,β-unsaturated carbonyls and nitriles such as methyl crotonate and crotononitrile are dimerized in high yields at 90-110° by the binary catalyst system formed by a copper compound and an isocyanide. Under the same conditions, β-unsubstituted α,β-unsaturated carbonyl and nitrile compounds such as

$$2RR'CHCH=CHX \xrightarrow{Cu_2O+C_6H_{11}NC} \begin{array}{c} RR'CHCH=CX \\ | \\ RR'CHCHCH_2X \end{array}$$

$$R = CH_3, \quad R' = H, \quad X = CN, \quad COOCH_3, \quad COCH_3$$
$$R = C_3H_7, \quad R' = H, \quad X = COOCH_3$$
$$R = CH_3, \quad R' = CH_3, \quad X = COOCH_3$$

acrylates and acrylonitrile are not dimerized, byt they are codimerized with β-alkyl-substituted monomers [67].

$$RR'CHCH=CHX+CH_2=CHY \xrightarrow{Cu_2O+C_6H_{11}NC} \begin{array}{c} RR'CHCH=CH \\ | \\ CH_2CH_2Y \end{array}$$

$$R = CH_3, \quad R' = H, \quad X, \quad Y = COOCH_3$$
$$R = CH_3, \quad R' = H, \quad X, \quad Y = CN$$

Of the copper compounds, cuprous oxide is the most active. Other copper compounds, e. g., copper(II) acetylacetonate, cupric oxide, and cuprous and cupric chlorides are less active. A fairly large amount of isocyanide is required for high catalyst activity.

Using the copper-isocyanide system, 3-butenenitrile is isomerized to 2-butenenitrile by double-bond migration during dimerization. The structure of the dimer corresponds exactly to that of the dimer of 2-butenenitrile. *Cis-trans* isomerization of the latter compounds also takes place during the dimerization. Furthermore, the exchange between the hydrogen at the α-position and that at the γ-position in 2-butenenitrile is caused by the catalyst system. A mechanism which is consistent with the above data was proposed as shown below [67b]:

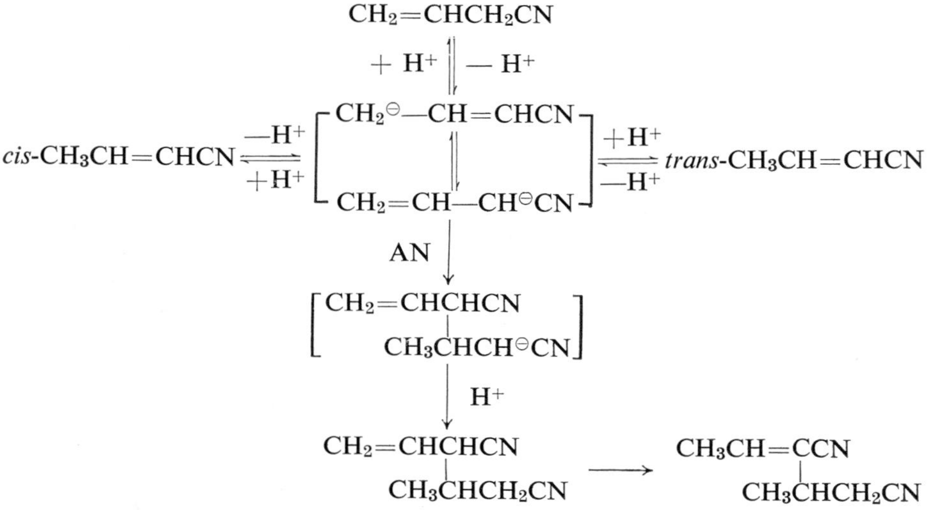

$$CH_2=CHCH_2CN$$

$$+ H^+ \updownarrow - H^+$$

$$cis\text{-}CH_3CH=CHCN \underset{+H^+}{\overset{-H^+}{\rightleftharpoons}} \left[\begin{array}{c} CH_2^{\ominus}—CH=CHCN \\ \updownarrow \\ CH_2=CH—CH^{\ominus}CN \end{array}\right] \underset{-H^+}{\overset{+H^+}{\rightleftharpoons}} trans\text{-}CH_3CH=CHCN$$

$$AN \downarrow$$

$$\left[\begin{array}{c} CH_2=CHCHCN \\ | \\ CH_3CHCH^{\ominus}CN \end{array}\right]$$

$$\downarrow H^+$$

$$\begin{array}{ccc} CH_2=CHCHCN & & CH_3CH=CCN \\ | & \longrightarrow & | \\ CH_3CHCH_2CN & & CH_3CHCH_2CN \end{array}$$

The allyl carbanion copper isocyanide complex was postulated as a key intermediate, which is formed by γ-hydrogen abstraction of *cis*- and *trans*-2-butenenitriles and by α-hydrogen abstraction of 3-butenenitrile. The addition of the carbanion to the second molecule of monomer followed by the double bond shift gives the final dimer.

In the presence of a protic solvent such as *t*-butyl alcohol acrylonitrile and acrylate are dimerized to 2-methyleneglutaronitrile and 2-methyleneglutarate, respectively, by the above binary catalyst system. Transition metal acetylaceto-nates, such as Fe(II and III), Co(II and III), and Ni(II), in combination with isocyanide are also effective for this dimerization [68]. A vinyl carbanion complex was assumed to be the key intermediate in the reaction since the deuterium/hydrogen exchange occurred at the α-position of acrylonitrile upon treatment with the copper acetylacetonate-cyclohexyl isocyanide system in the presence of *t*-butanol-d.

$$CH_2=CHX \; \underset{}{\overset{e}{\rightleftharpoons}} \; [CH_2=C^{\ominus}-X] \; \xrightarrow{CH_2=CHX} \; \begin{bmatrix} CH_2=CX \\ | \\ CH_2CH^{\ominus}X \end{bmatrix}$$

$$\begin{array}{ccc} & CH_2=CHX & \downarrow H^+ \\ -\begin{pmatrix} CH_2-CH \\ | \\ X \end{pmatrix}_n & & \begin{array}{c} CH_2=CX \\ | \\ CH_2CH_2X \end{array} \end{array}$$

$$X = CN, \quad COOCH_3$$

Recently several catalyst systems of new types for dimerizing acrylonitrile have been claimed in patents. For instance, in the presence of CuCl and a base such as triethylamine or triphenylphosphite acrylonitrile was dimerized under a hydrogen atmosphere into a mixture of 1,4-dicyanobutenes and adiponitrile [69]. Use of a binary catalyst system composed of a transition metal compound such as an iron, copper, nickel, or cobalt salt and an organoaluminum compound, especially triisobutylaluminum, gave the straight chain head-to-head dimers from α,β-unsaturated esters, amides, and nitriles. Thus, by the catalyst system composed of copper(II) acetylacetonate and triisobutylaluminum acrylonitrile was dimerized at 10° into 1,4-dicyano-2-butene [70]. Adiponitrile was prepared by reductive dimerization of acrylonitrile by using iron, chromium, manganese, or cobalt catalysts supported on Al_2O_3 or activated charcoal at 60-120° and 50-300 atm of hydrogen [71].

2-Methyleneglutaronitrile was catalytically obtained at ambient temperature by the use of a binary system composed of a transition metal halide and trialkylamine [72]. Besides the dimer, a trace of a trimer, 2, 4, 6-tricyanohexene-1 is formed. The order of increasing catalytic activity is Co(II) $\geqslant$ Zn(II)

$> Fe(II) > Al(III) > V(III) > Ti(IV)$. Triethylamine is especially effective in the reaction:

$$CH_2=CHCN \quad \xrightarrow[20-30^\circ]{MX_n+NR_1R_2R_3}$$

$$\underset{NCCCH_2CH_2CN}{\overset{CH_2}{\parallel}} \quad + \quad \underset{\underset{CH_2CH_2CN}{\overset{|}{}}}{\underset{NCCCH_2CHCN}{\overset{CH_2}{\parallel}}}$$

The condensation of two molecules of acetylene with one molecule of an acrylic compound takes place in the presence of nickel complexes [73]. The interaction of acetylene and methyl acrylate gives in high yield methyl 2,4,6-heptatrienoate. The reaction also proceeds readily with acrylonitrile to form 2,4,6-heptatrienenitrile. The catalyst is prepared by adding nickel carbonyl

$$2CH \equiv CH + CH_2=CHX \longrightarrow CH_2=CH-CH=CH-CH=CH-X$$

$$X = CN, COOCH_3$$

and triphenylphosphine to the acrylic compound. Other nickel compounds, such as nickel cyanide, in combination with triethylphosphite, are also active as catalysts.

Acrylates are codimerized with 1,3-dienes by the use of the binary catalyst system consisting of cobalt(III) or iron (III) acetylacetonate and triethylaluminum [74]:

$$CH_2=CR-CH=CH_2 + CH_2=CH-COOR' \longrightarrow$$

$$CH_3-CH=CR-CH_2-CH=CH-COOR' +$$

$$+ CH_3-CR=CH-CH_2-CH=CH-COOR'$$

$$R = H \text{ or } CH_3$$

$$R' = CH_3, C_2H_5, \text{ or } n\text{-}C_4H_9$$

3-Methylhepta-1,4,6-triene, which is the main product of the butadiene dimerization using the cobalt catalyst systems, was not detected. In both cobalt and iron catalyst systems, the activity of the catalyst was markedly suppressed when the catalyst solution was prepared in the presence of methyl acrylate without butadiene.

Methyl acrylate and butadiene give a good yield of methyl-2,5,10-undecatrienoate at 80-120° with the catalyst system composed of nickel(II) acetylacetonate, diethylaluminum ethoxide, and triphenylphosphine [75].

$$2CH_2=CH-CH=CH_2 + CH_2=CH-COOCH_3 \longrightarrow$$

$$\longrightarrow CH_2=CH-CH_2-CH_2-CH_2-CH=CH-CH_2-CH=CH-COOCH_3$$

6. CONCLUSION

Dimerization and co-dimerization of acrylic compounds have been reviewed. The bifunctional products obtained in these reactions are important industrial intermediates, for this reason the achievement of the inexpensive dimerization of readily available acrylic compounds has been sought after in many laboratories in recent years. Several types of reactions reported in the literature and patents, particularly those in which reactions have been catalyzed by transition metal complexes, have been described in more detail in this review.

Coordination catalysis has opened a very attractive way to new products by dimerization of unsaturated monomers. This method has now been broadened to include acrylic compounds and several transition metal complexes have been found as catalysts. It is hoped that in the near future, many more catalysts will be found in this field and industrial processes to obtain bifunctional materials will be developed.

However this type of reactions may be of great interest also in laboratory scale and may be considered in detail by organic chemists.

7. REFERENCES

[1] a) Y. WATANABE and M. TAKEDA, "Shokubai", 10, 104, (1968).

    b) Y. UCHIDA and M. HIDAI, "Kagaku no Ryoiki Zokan 89", p. 139, (1970), Nankodo, (Tokyo).

    c) G. LEFÉBVRE and Y. CHAUVIN, "Aspects Homog. Catalysis", 1, 107, (1970), Ed. R. Ugo, Manfredi, (Milan).

[2] a) K. ALDER, H. OFFERMANNS, and E. RUEDEN, "Chem. Ber.", 74B, 905, (1941).

    b) K. ALDER and E. RUEDEN, "Chem. Ber.", 74B, 920, (1941).

[3] E. C. COYNER and W. S. HILLMAN, "J. Amer. Chem. Soc.", 71, 324, (1949).

[4] Brit. Pat. (to E. I. Du Pont de Nemours and Co)., 897, 275 (1965).

[5] a) Jap. Pat. (to Knapsack-Griesheim A. G.), 63 1, 643.

    b) Jap. Pat. (to Knapsack-Griesheim A. G.), 64 12,622.

    c) Jap. Pat. (to Mitsui Petrochemical Co.), 68 22,979.

    d) Neth. Appl. (to Badische Anilin und Soda-Fabrik), 6,516,357 (1966).

[6] a) J. RUNGE and R. KACHE, "Z. Chem.", 8 (10), 382b-3, (1968); French Pat. (to Veb. Chemische Werke Buna), 1,470,282 (1967).

    b) S. HOSAKA and S. WAKAMATSU, "Tetrahedron Lett.", 219 (1968); Jap. Pat. (to Toyo Rayon Co.), 68 2,699.

    c) R. S. H. LIU and D. M. GALE, "J. Amer. Chem. Soc.", 90, 1897, (1968).

[7] a) GG. SPIEGELBERGER and O. BAYER, Ger. Pat., 71,937 (1942).

    b) P. B. Report, 20,545.

    c) U. S. Pat. (to National Distillers and Chemical Co.), 3,133,956 (1964).

[8] R. M. LEEKLEY, U. S. Pat., 2,439,308 (1948).

[9] *a*) I. L. KNUNYANTS et al., U. S. S. R. Pat., 105, 286 (1954).

   b) I. L. KNUNYANTS, "Izvest. Akad. Nauk SSSR, Otdel Khim. Nauk", No. 2,238, (1957).

   c) I. L. KNUNYANTS, "Doklady Akad. Nauk, Otdel Khim. Nauk", *113*, 112 (1957).

   d) I. L. KNUNYANTS, *Soveshchanie Po Electrokhimii, 4th Moscow*, 1956; "Chem. Abst.", *54*, 9811*b* (1960).

[10] M. M. BAIZER, "J. Electrochem. Soc.", *111*, 215, (1964).

[11] M. M. BAIZER and J. D. ANDERSON, "J. Electrochem. Soc.", *111*, 223,226, (1964).

[12] *a*) M. M. BAIZER, "Tetrahedron Lett.", 973, (1963).

   b) M. M. BAIZER, "J. Org. Chem.", *29*, 1670, (1964).

[13] Y. ARAD, M. LENY, I. R. MILLER, and D. VOFSI, "J. Electrochem. Soc.", *114*, 899, (1967).

[14] F. MATSUDA, "Tetrahedron Lett.", *49*, 6193, (1966).

[15] Australian Pat. (to Imperial Chemical Industries Ltd.), 662, 661 (1964).

[16] *a*) M. FIGEYS, "Tetrahedron Lett.", 2237 (1967).

   b) M. FIGEYS and H. D. FIGEYS, "Tetrahedron", *24*, 1097, (1968).

[17] N. TAKASHIMA and C. C. PRICE, "J. Amer. Chem. Soc.", *84*, 489, (1962).

[18] U. S. Pat. (to American Cyanamid), 3,074,999 (1963).

[19] M. M. BAIZER and J. D. ANDERSON, "J. Org. Chem.", *30*, 1357, (1965).

[20] W. H. DIETSCHE, "Tetrahedron Lett.", 6347, (1966).

[21] *a*) French Pat. (to Société des Usines Chimiques Rhône-Poulenc), 1,366,081, (1964).

   b) Brit. Pat. (to Société des Usines Chimiques Rhône-Poulenc), 1,003,656, (1964).

[22] *a*) U. S. Pat. (to Shell Oil Co.), 3,225,083, (1965).

   b) J. D. MCCLURE, "J. Org. Chem.", *35*, 3045, (1970).

[23] R. ODA, T. KAWABATA, and S. TANIMOTA, "Tetrahedron Lett.", 1653, (1964).

[24] J. D. MCCLURE, "Tetrahedron Lett.", 2401, (1967).

[25] H. MUELLER, D. WITTENBERG, and D. MANGOLD, "Ind. Chim. Belge", 32 (Spec. No. Pt. III), 38, (1967).

[26] *a*) French Pat. (to Badische Anilin und Soda-Fabrik A. G.), 1,385,883, (1965).

   b) French Pat. (to National Distillers and Chemical Co.), 1,388,744, (1965).

   c) Neth. Appl. (to National Distillers and Chemical Co.), 66,02627, (1966).

   d) Brit. Pat. (to Imperial Chemical Industries Ltd.), 107,655, (1967).

   e) French Pat. (to Badische Anilin und Soda-Fabrik A. G.), 1,522,836, (1968).

   f) French Pat. (to E. I. Du Pont de Nemours and Co.), 1,519,376, (1968).

[27] U. S. Pat. (to Halcon International, Inc.), 3,538,141, (1970).

[28] *a*) Neth. Appl. (to E. I. Du Pont de Nemours and Co.), 67,05504, (1967).

   b) Neth. Appl. (to Röhm and Haas Co.), 67,01120, (1967).

   c) French Pat. (to Röhm and Haas Co.), 1,499,708, (1967).

   d) Jap. Pat. (to Mitsubishi Rayon Co.), 70 35,288.

[29] U. S. Pat. (to Eastman Kodak Co.), 2,675,372, (1954).

[30] U. S. Pat. (to Shell Oil Co.), 3,227,7 5, (1966).

[31] Brit. Pat. (to Imperial Chemical Industries Ltd.), 1,100,350, (1968).

[32] *a*) Jap. Pat. (to Röhm and Haas Co.), 66 19,331.
  b) Ger. Offen. (to Toa Gosei Chemical Industry Co., Ltd.), 1,904,361, (1969).

[33] Brit. Pat. (to Imperial Chemical Industries Ltd.), 1,128,320, (1968).

[34] K. MORITA and T. KOBAYASHI, "Bull. Chem. Soc. Japan", *42*, 2732, (1969).

[35] W. REPPE, O. SCHLICHTING, K. KLAGER, and T. TOPEL, "Annalen", *560*, 1, (1968).

[36] R. CRAMER, "J. Amer. Chem. Soc.", *87*, 4717, (1965).

[37] E. WEISS, K. STARK, J. E. LANCASTER, and H. D. MURDOCH, "Helv. Chim. Acta", *46*, 288, (1963).

[38] *a*) G. N. SCHRAUZER, "J. Amer. Chem. Soc.", *81*, 5310, (1969).
  b) G. N. SCHRAUZER, "Chem. Ber.", *94*, 642, 650, (1961).

[39] B. L. ROSS, J. G. GRASSERI, W. M. RITCHEY, and H. D. KAESZ, "Inorg. Chem.", *2*, 1023, (1963).

[40] R. J. KERN, "J. Inorg. Nucl. Chem.", *25*, 5, (1963).

[41] S. F. A. KETTLE and L. E. ORGEL, "Chem. and Ind. (London)", 49, (1960).

[42] J. F. GUTTENBERGER and W. STROHMEIER, "Chem. Ber.", *100*, 2807, (1967).

[43] French Pat. (to Société des Usines Chimiques Rhône-Poulenc), 1,377,425, (1964).

[44] French Pat. (to Société des Usines Chimiques Rhône-Poulenc), 1,381,511, (1964).

[45] U. S. Pat. (to E. I. Du Pont de Nemours and Co.), 3,206,498, (1965).

[46] U. S. Pat. (to E. I. Du Pont de Nemours and Co.), 3,322,819, (1967).

[47] A. MISONO, Y. UCHIDA, K. TAMAI, and M. HIDAI, "Bull. Chem. Soc. Japan", *40*, 931, (1967).

[48] P. KRUMHOLZ and H. M. A. STETTINER, "J. Amer. Chem. Soc.", *71*, 3035, (1949).

[49] A. MISONO, Y. UCHIDA, M. HIDAI, and H. KANAI, "Bull. Chem. Soc. Japan", *40*, 2089, (1967).

[50] French Pat. (to Société des Usines Chimiques Rhône-Poulenc), 1,453,988, (1966).

[51] *a*) French Pat. (to Toyo Rayon Co.), 1,542,199, (1966).
  b) French Pat. (to Toyo Rayon Co.), 1,524,921, (1967).

[52] T. ALDERSON, E. L. JENNER, and R. V. LINDSEY, Jr., "J. Amer. Chem. Soc.", *87*, 5638, (1965).

[53] K. C. DEWHIRST, "Inorg. Chem.", *5*, 319, (1966).

[54] *a*) Neth. Appl. (to Société des Usines Chimiques Rhône-Poulenc), 66,03115, (1966).
  b) French Pat. (to Société des Usines Chimiques Rhône-Poulenc), 1,451,443, (1966).
  c) French Pat. (to Société des Usines Chimiques Rhône-Poulenc), 1,493,068, (1968).
  d) Brit. Pat. (to Société des Usines Chimiques Rhône-Poulenc), 1,102,460, (1968).
  e) French Pat. (to Société des Usines Chimiques Rhône-Poulenc), 1,546,530, (1968).

[55] *a*) A. MISONO, Y. UCHIDA, M. HIDAI, and H. KANAI, "Chem. Commun.", 354, (1967).
  b) A. MISONO, Y. UCHIDA, M. HIDAI, H. SHINOHARA, and Y. WATANABE, "Bull. Chem. Soc. Japan", *41*, 396, (1968).

c) A. MISONO, Y. UCHIDA, M. HIDAI, and I. INOMATA, "Chem. Commun.", 704, (1968).

d) A. MISONO, Y. UCHIDA, M. HIDAI, I. INOMATA, Y. WATANABE, and M. TAKEDA, "Kogyo Kagaku Zasshi", *72*, 1801, (1969).

[56] *a*) French Pat. (to Mitsubishi Petrochemical Co.), 1,560,994, (1969).

    b) Brit. Pat. (to Mitsubishi Petrochemical Co.), 1,177,059, (1970).

[57] *a*) Brit. Pat. (to E. I. Du Pont de Nemours and Co.), 1,133,900, (1968).

    b) French Pat. (to E. I. Du Pont de Nemours and Co.), 1,572,892, (1969).

    c) Ger. Offen. (to Kurashiki Rayon Co.), 1,903,657, (1969).

    d) Jap. Pat. (to Kurashiki Rayon Co.), 71 06,891.

[58] Neth. Appl. (to Halcon International, Inc.), 68,03864, (1967).

[59] J. D. MCCLURE, R. OWYANG, and L. H. SLAUGH, "J. Organometal. Chem.", *12*, p. 8, (1968).

[60] E. BILLIG, C. B. STROW, and R. L. PRUETT, "Chem. Commun.", 1307, (1968).

[61] French Pat. (to Imperial Chemical Industries Ltd.), 1,519,113, (1966).

[62] Y. KOBAYASHI and S. TAIRA, "Tetrahedron", *24*, 5763, (1968).

[63] M. G. BARLOW, M. J. BRYANT, R. N. HASZELDINE, and A. G. MACKIE, "J. Organometal. Chem.", *21*, 215, (1970).

[64] J. KWIATEK, I. L. MADOR and J. K. SEYLER, "Advan. Chem. Ser.", *37*, 201, (1963).

[65] *a*) G. AGNES, G. P. CHIUSOLI, and G. COMETTI, "Chem. Commun.", 1515, (1968).

    b) G. AGNES, G. P. CHIUSOLI, and G. COMETTI, *IVth Int. Conference on Organometallic Chemistry*, Bristol (1969).

[66] J. SHABTAI and H. PINES, "J. Org. Chem.", *30*, 3854, (1965).

[67] *a*) T. SAEGUSA, Y. ITO, S. KOBAYASHI, and S. TOMITA, "Chem. Commun.", 273, (1968).

    b) T. SAEGUSA, Y. ITO, S. TOMITA, and H. KINOSHITA, "J. Org. Chem.", *35*, 670, (1970).

[68] T. SAEGUSA, Y. ITO, H. KINOSHITA, and S. TOMITA, "Bull. Chem. Soc. Japan", *43*, 877, (1970).

[69] French Pat. (to Imperial Chemical Industries Ltd.), 1,519,114, (1968).

[70] French Pat. (to Imperial Chemical Industries Ltd.), 1,527,980 (1968).

[71] Ger. Pat. (to Metallgesellschaft A. G.), 1,270,026, (1968).

[72] Ger. Offen. (to Mitsubishi Petrochemical Co.), 1,922,017, (1969).

[73] T. L. CAIRNS, V. A. ENGELHARDT, H. L. JACKSON, G. H. KALB, and J. C. SAUER, "J. Amer. Chem. Soc.", *74*, 5636, (1952).

[74] *a*) H. MÜLLER, D. WITTENBERG, H. SEIBT, and E. SCHARF, "Angew. Chem.", *77*, 318, (1965).

    b) A. MISONO, Y. UCHIDA, T. SAITO, and K. UCHIDA, "Bull. Chem. Soc. Japan", *40*, 1889, (1967).

[75] French Pat. (to Société des Usines Chimiques Rhône-Poulenc), 1,433,409, (1966).

Author index